BRAIN-BODY PARENTING

ALSO BY MONA DELAHOOKE, PHD

Beyond Behaviors: Using Brain Science and Compassion to Understand and Solve Children's Behavioral Challenges

Social and Emotional Development in Early Intervention

BRAIN-BODY PARENTING

HOW TO STOP MANAGING BEHAVIOR AND START RAISING JOYFUL, RESILIENT KIDS

MONA DELAHOOKE, PHD

HARPER WAVE
An Imprint of HarperCollins*Publishers*

The information in this book is intended to be a source of information only. This book contains advice and information which should be used to supplement rather than replace the advice of your child's doctor or another trained pediatric professional. It is recommended that you seek advice from a trained professional before embarking on any treatment program. The author and publisher disclaim responsibility for any adverse effects arising from the application of the information contained herein.

All identifying information, including names and other details, has been changed to protect the privacy of individuals. Any similarity to actual individuals or families is coincidental.

HarperCollins books may be purchased for educational, business, or sales promotional use. For information, please email the Special Markets Department at SPsales@harpercollins.com.

FIRST EDITION

Designed by Elina Cohen
Title page art courtesy of Shutterstock/Valenty
Part I art courtesy of Shutterstock/Singleline
Part II art courtesy of Shutterstock/OneLineStock.com

Library of Congress Cataloging-in-Publication Data has been applied for.

ISBN 978-0-06-306131-6

22 23 24 25 26 LSC 10 9 8 7 6 5 4 3 2 1

To my mother, Clara, whose love
built the foundation of my very being

CONTENTS

INTRODUCTION:

Personalizing Your Parenting

As soon as I heard Janine's voice on the phone, I could tell she was distraught.

She had lost her cool with her four-year-old son, Julian. What had begun as a routine shopping trip had spun out of control, morphing into a disaster in the parking lot of the local Target store and leaving Janine feeling both angry and embarrassed.

The next morning, sitting on a couch in my office, she was still raw with emotion as she recounted the incident. Janine was hardly a novice at managing children. She had taught second grade for a decade before becoming a mom, displaying such classroom prowess that she had won multiple teaching awards. As for Julian, he had been a precocious child who walked at eleven months and was speaking a few words by his first birthday. But he sometimes had difficulty controlling his emotions and following instructions. Even when his mother offered incentives for good behavior, Julian often struggled to comply.

That day, as mother and son were waiting in the checkout line, Julian had suddenly grabbed a candy bar from the rack.

"Put that back, please!" Janine pleaded.

When Julian refused, Janine shifted instinctively into the mode that had served her well handling classroom outbursts: First she tried to distract her son, and then, when that failed, she told him that if he didn't relinquish the candy, she wouldn't give him the reward sticker she routinely placed on his behavior chart each day. Janine did her best to be consistent and calm. But on this day, the more she spoke to her son, the more defiant he became.

Finally, Julian snapped, hurling the candy bar smack in the cashier's face.

"Julian!" Janine screamed. "You are being *so bad*!" She quickly apologized to the clerk and then abandoned her shopping cart to carry her son, now bawling, to the parking lot. There, she shoved him into his car seat and then she, too, broke into tears.

"I didn't recognize myself," Janine told me the next day. "I just feel so guilty." She had called her son "bad" when everything she knew about him told her otherwise. When he was calm, he was loving, gentle, and polite.

Why hadn't Janine's efforts succeeded in calming her son? Why had things gone so quickly and horribly wrong? And what could she have done differently? Nearly everyone who raises or cares for children has experienced plenty of moments like Janine's. In more than three decades as a child psychologist, I have met countless parents like her: caring, compassionate, insightful people, eager to see their children flourish and thrive, but now wondering what they are missing. Again and again, I hear the same refrain: *We're doing what the parenting books say to do! Why is it falling short?*

These parents are asking the same questions mothers and fathers have posed for generations—questions you may have, too: Why does my daughter refuse to cooperate or listen to me? Why is my son's behavior so unpredictable? Why is he such a picky eater? Why can't she sleep through the night? How can we set appropriate limits? How can I tell whether I'm expecting too much from my child—or too little? And then there's perhaps the most important question Janine asked me that morning, one that I've heard from countless parents before and since: Why do I keep losing it with my child when I know better?

Just as they share these questions and concerns, most parents also share the same deeply felt desire: to raise children who will grow up to be resilient, confident, happy, and independent people. But how to do that? Whose parenting advice is worth following? Parents today are bombarded with more guidance from more perspectives than ever before, from social media influencers to friendly (or judgmental) neighbors to school specialists to TED Talks to whatever a Google search turns up. You can choose among conscious parenting and attachment parenting and free-range parenting and dozens of other philosophies. Which is best? And what's the optimal solution to challenging behaviors? Time-outs? Reasoning? Ignoring? Pausing and counting to three? Something else?

Of course, parents want what's best for their children, but many of those I see in my practice are understandably confused and perplexed about what that is. With so many perspectives to choose from, whose wisdom can you trust? In more than three decades working with children and their families, I have come to realize that there is no one-size-fits-all approach to successful child-rearing. What's most important isn't the rules but the child. *What's crucial isn't understanding someone else's guidelines but understanding how our parenting is "landing" in our child.* Once we have some insight into *how* a child is absorbing the interactions and circumstances they experience, we can begin to discover more personalized and more effective answers to common parenting questions.

The problem is that too often we focus on a child's *behaviors* instead of the child. We're concerned about *solving problems* instead of *cultivating relationships and building bonds.*

In this book, I'll share how you can shift the focus from the behavior to what's *underlying* the behavior; to redirect your efforts from understanding a parenting approach to understanding your child. My aim is to help you adapt your parenting to the individual needs of your child and, in turn, build a connection that helps them become more resilient.

In short, parenting isn't about a theory or a hypothetical approach. It's about you and your child. This book will help you shift from managing behaviors to using those behaviors as clues to help understand your child's inner reality—the child's sensory experiences, feelings, and emotions.

Most parenting books suggest "top-down" responses to behavior that are oriented to the child's brain: talking, reasoning, incentivizing, or offering rewards or consequences. These approaches generally offer parents two options: Reason with the child or discipline the child. While these approaches acknowledge a child's cognitive (thinking) capacity, they neglect to account for the child's entire nervous system; in other words, the brain-body connection. After all, the nervous system runs throughout the body, and it sends feedback to the brain. This book acknowledges the equal importance of the *body and brain* to understanding the child. It will show you how to parent using not just *psychology* but also *biology*.

It's important to note up front that the approach you'll learn in these pages isn't just based on my own thoughts or observations. It's grounded in cutting-edge neuroscience. It is also a product of my experience as both a clinical psychologist and a mother.

As a graduate student in psychology, I learned to focus on recognizing and diagnosing what was wrong—assigning labels to psychological symptoms. When I had my own child, I discovered that the top-down approaches I had learned in grad school, the methods that called for appealing to the child's mind, didn't always work. At least, they didn't help me figure out how to help *my* baby, who cried for hours on end, or how to convince my anxious ten-year-old that yes, she *could* handle a sleepover at a friend's house. So a decade into my career, I stepped away from my practice to seek out better answers to the challenges that parents brought to me and that I faced myself as a mother. What I discovered profoundly changed my practice and my view of parenting.

I decided to start from the beginning. My traditional training hadn't included much insight into infant development and how to help babies, so I enrolled in two training programs focused on infant mental health. I spent three years in hospital, clinic, and preschool settings, studying babies and toddlers. The experience opened my eyes to the crucial impact of early development.

I also learned how the *individual differences within each child's body* influenced the child's development and how their parents and other adults interacted with them. Working on multidisciplinary teams that included pediatricians, speech and language therapists, physical therapists, occupational therapists, educators, psychologists, and—most important—parents, I learned how a baby's *body-based* interpretation of the world affected development and early relationships.

My traditional psychology training had focused on top-down approaches anchored in the child's brain. But the only way infants communicate is through their bodies. Understanding the "bottom-up" (or "body-up") experiences that precede the development of the child's thinking and formation of concepts gave me a better way to understand children at all stages of development.

As I continued to explore and learn, I had the good fortune to study with two pioneers in early child development, Dr. Stanley Greenspan, a psychiatrist, and Dr. Serena Wieder, a psychologist, whose clinical model was among the first to incorporate the brain and body in early intervention. They began with the premise that it's essential to help a child calm down—to regulate the *body*—before talking, reasoning, or offering incentives can succeed.

More importantly, they based their model on the idea that the only way that humans successfully regulate their bodies is through attuned, loving, and safe relationships. That explained why the top-down approaches of my psychology training often failed. My training had overlooked the essential role of feedback from the body and its profound impact on *relationships* and how they, in turn, affect children's behaviors. At around the same time (this was the 1990s), scientists were learning new information about the human brain at such an explosive rate that this period came to be known in scientific circles as the "decade of the brain."

As it turned out, what I was learning about the centrality of relationships gained support and validation from the emerging field of relational neuroscience. Dr. Dan Siegel, a psychiatrist, founded the field of interpersonal neurobiology, which studies the impact of interpersonal experiences on brain development. And Dr. Bruce Perry, another psychiatrist, formed the neurosequential model of therapeutics (NMT), which echoed what Drs. Greenspan and Wieder had taught me about how relationships promote a calm, "regulated" body, which is essential to the capacity to learn and grow. Dr. Connie Lillas—a nurse, marriage and family therapist, and researcher—co-developed the NRF, the Neurorelational Framework, which similarly highlights the central importance of relationships that influence brain development. More recently, I have studied the work of Dr. Lisa Feldman Barrett, a neuroscientist whose theory of constructed emotion emphasizes the impact of deep-body signals on our basic feelings and emotions.

It was a neuroscientist, Dr. Stephen Porges, whose groundbreaking work changed me most as a professional and a parent. His Polyvagal theory, first introduced in 1994, offered an elegant explanation from human evolution for how and why humans react to life's various circumstances. Dr. Porges's work provided a new understanding of how the autonomic nervous system—the great information highway connecting the brain and the body—influences human physiology, emotions, and behaviors.

Dr. Porges's work provided a neuroscience theory base that

brought the body and brain together into a fuller understanding of children's behaviors. The Polyvagal theory offered the scientific reasoning for all I had learned about the critical importance of loving and attuned relationships that are personalized to each child's individual differences and nervous system. It was reassuring to discover this perspective, which stood in contrast with my original training in behavior management. What I have learned from these scientists and therapists formed this book's core message: that regulation in a child's physical body supports healthy relationships and loving interactions, in turn building the infrastructure that eventually enables the child to use reasoning, concepts, and thinking to flexibly manage life's challenges. With this understanding of the two-way communication between the brain and body, I shifted my practice from focusing on eliminating children's disruptive behaviors to understanding them as the body's way of communicating its needs.

I chronicled that shift in my 2019 book *Beyond Behaviors*, which called for a change in how we respond to children's challenging behaviors, and introduced a new way of supporting them. That book struck a chord with many people, and I've heard from parents, teachers, mental-health professionals, and therapists all over the world who resonated with its call for a paradigm shift in education and psychology. That consensus coincided with the rise of the neurodiversity movement and its calls to honor and respect individual differences rather than pathologizing them as disorders that need to be fixed. The new paradigm I was advocating recognized that the central and sacred ingredient of all child development is the relationship between parent and child.

At this exciting moment, *Brain-Body Parenting* builds on those lessons, adding the important insight that the brain-body connection provides a new foundation for understanding children's behaviors and leads to a new parenting road map for nurturing joyful and resilient kids.

What you will find in this book isn't a definitive guide to the neuroscience research and theories, but rather what I have taken from the emerging science and how I applied it in real life as a psychologist and as a mother. These chapters contain my own translation of the science for practical use. As such, my descriptions *only begin* to scratch the surface of the complexity of the emerging science, and I have taken great liberty in simplifying concepts to make them accessible. Neuroscience research is evolving rapidly, so stay tuned for updates. In the back of the book, you'll find a glossary that includes some of the significant scientific concepts.

Science shouldn't be confined to laboratories and medical journals. We should bring it into our kitchens and living rooms, where it can reduce suffering, improve relationships, and guide our everyday parenting decisions. (If you're so inclined, I encourage you to read the original sources, cited in the bibliography and endnotes.)

With the support of neuroscience, instead of focusing on helping parents change their child's behaviors, I began working with children and parents together, helping parents to understand their children—and themselves—in a more holistic way, beginning with an appreciation of each child's brain-body connection.

When we recognize that a child's brain doesn't operate in isolation from the child's body, a new array of parenting options emerges. Understanding the brain-body connection that underlies all behavior gives us a new road map to guide our parenting decisions, one that is tailored to each child. For more than two decades, I have used this idea to help parents understand and solve common parenting dilemmas.

Instead of seeking a psychological diagnosis, we'll seek to understand the child's physiology that is contributing to the behaviors. Instead of looking for deficits, we'll listen to the body's signals to detect clues. Seeing your child's behaviors, attitudes, and actions through the lens of the child's nervous system will help you *personalize* your parenting, giving you a road map for making your important parenting decisions.

That's what I began to share with Janine, Julian's mom, the morning when she was so frustrated with herself, exasperated that nothing could calm her son—or herself—after the Target blowup. It wasn't her fault that her teacher training hadn't covered the impact that stress can have on the nervous system. Nor could she be faulted for Julian's pediatrician blaming the boy's difficulty managing disappointment on his "strong-willed" nature and encouraging her to use rewards and consequences to stop "bad" behaviors and promote compliance. Indeed, many of my colleagues in mental health and education still recommend similar approaches.

What I discussed with Janine was that her son wasn't intentionally testing her limits or being willfully uncooperative; he was responding to stress. The answer in such situations isn't to discipline the child or offer the child incentives but instead to customize your parenting, to help the child calm their nervous system so they can function, engage, and learn. That's the foundation of the approach I shared with her—one that is compassionate, holistic, and effective. It's the approach I'll share in this book.

In the pages to come, we'll learn key concepts to personalize your parenting, including:

- *How to use your child's behaviors as clues to learn about your child's unique* platform—*my simplified term for the brain-body connection.*
- *How the concept of* neuroception, *Dr. Porges's term for what I call the body's safety-detection system, helps us understand our child's subjective experiences.*
- *How to discern whether a given behavior stems from willful intention or physiological (body-based) distress.*
- *How a shared connection between parent and child can help a child develop the ability to self-regulate.*

- *The importance of self-care for parents, and how the loving presence of an adult can help a child to feel calmer and safer physiologically.*
- *How understanding* interoception—*the sensations coming from deep inside the body—can help guide your interactions, and help your child become more aware of and conversant with feelings and emotions.*
- *How to apply an understanding of the platform and all of these concepts to help build resilience and solve common parenting challenges from infancy to early adolescence.*

RELIEVING, NOT ADDING TO, THE WORK OF PARENTING

You might be thinking: *Okay, but that sounds like a lot of work.* Before we move on, let me assure you that my aim is not to add to your burden as a parent but rather to ease it. I never want to add stress, pressure, anxiety, or work. Far from it. This book isn't about becoming some kind of super-parent. I have worked with many hundreds of parents, and my belief is that parents do the best they can with the information they have. I'll help shine a light on all the knowledge you can glean from observing your child in a new way. Nobody knows your child as well as you do, and the tools and perspectives you'll find in this book are meant to help you build your relationship with your child in a natural and wonderful way. Our aim isn't to change your child but to help you personalize your relationship in a way that is grounded in deeply respecting your child's individual, ever-changing brain and body.

Finally, I would be remiss if I didn't acknowledge the importance of honoring your own parenting values, influenced by your unique background and life experiences. I was raised in the United

States by parents from two different continents. As a firstborn child and a first-generation American, I am a product of both their heritages, and the values and traditions they passed down to me were influenced by their cultures. Similarly, I expect that you will tailor the information in this book to your family's traditions and cultural values. Whether you are the parent of a baby or a much older child, a parent of one or a parent of many, I hope that the information you read here will empower you to grow in confidence as you view your child through a more comprehensive lens. It's not easy to be a parent, and I have built into this book important messages about how to exercise self-compassion and take care of yourself along the journey.

The key to understanding what underlies behaviors for children and parents alike is an appreciation for the brain *and* body. This knowledge changed me as a mother and as a psychologist, and when I began sharing it with the families I work with, it transformed their lives, too. I hope this book will help you to experience less worry and more joy as a parent, that it will lead to less second-guessing, self-judgment, and stress about your parenting decisions while nurturing your child's resilience and a connection that lasts a lifetime.

PART I

UNDERSTANDING BRAIN-BODY PARENTING

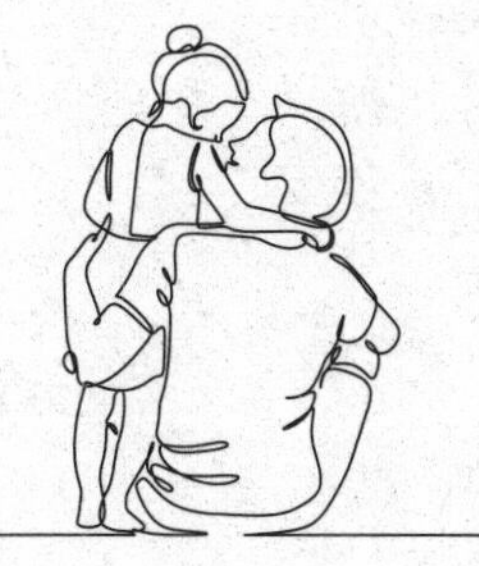

1

How to Understand Your Child's Physiology—and Why It's Important

We are the caretakers of each other's nervous systems as much as we are the caretakers for our own.

—Dr. Lisa Feldman Barrett

When Leanda and Ross found their way to my office, they were desperate for advice about their daughter Jade. They certainly seemed well equipped to be parents. Leanda was a pediatric nurse and Ross was a high school principal. Both had studied child development and read their share of parenting books. Their older daughter Maria's first years had gone relatively smoothly, and Jade, too, was happy and well adjusted early on.

Then came kindergarten. From the very first days, Jade put up a struggle. Each morning, as her father tried to drop her off at school, Jade begged Ross not to leave, screaming with such volume that ultimately a teacher would intervene, physically prying the girl from her dad.

The longer the daily struggle persisted, the more troubled and perplexed Jade's parents became. Was their daughter's resistance a sign of a serious problem? Would she grow out of it? Should they continue to follow the advice of the teachers, who advised them to drop off Jade and leave even if she protested—assuring them that this was a phase, something common in kindergartners? Or would another strategy be more beneficial? Mostly, they were perplexed about this question: Why was the parenting approach that had worked so well in raising their older daughter falling short with Jade?

By the time I met them, this daily battle had persisted for nearly four months, despite their nurturing parenting style, their open communication, and all the advice they had gathered from various parenting books. Leanda and Ross were clearly eager to see their child flourish, but they were stressed out and confused about what to do. How could they help Jade thrive?

The guidance they had been receiving from Jade's teachers was in line with common parenting advice: to look at the *behavior*, not the *child*. Most parenting approaches focus not on the whole child, but on the child's behaviors—and how parents should respond to particular kinds of behavior. And they suggest responses to behavior that are oriented to the child's brain: reasoning, requesting, or offering incentives, rewards, or consequences.

These reactive approaches have two inherent flaws. First, they offer one-size-fits-all answers, based not on *your* child but on a generic version of a child. And second, they assume that the child is intentionally behaving in a certain way—that in any given moment, the child is in control, or *if she tries hard enough*, can gain control of herself.

The specific advice is often guided by a particular philosophy: Be positive (focus on encouragement), or be supportive yet strong (authoritative), or let your child fail more (don't helicopter), or don't project your own issues onto your child (become more conscious), or reflect on present experiences without judgment (become more mindful), or help your child learn to talk about feelings (teach them about emotions). All of these concepts can be useful, but they fail to account for your child's unique traits or their needs at any particular moment. In other words, it doesn't matter how good the advice is in theory if the child isn't yet receptive to being taught.

And then there are the common misconceptions about children's willful or intentional control over their behavior. Not long ago, I watched a video online that claimed it would explain how to prevent toddlers from "throwing" tantrums. It had racked up more than a million views on YouTube. There's just one problem: Contrary to popular myth, toddlers don't generally have tantrums *on purpose*. Rather, tantrums *at any age* are a signal that the brain-body connection is in a state of overwhelm, challenge, or vulnerability.

PUTTING BEHAVIORS IN CONTEXT

When well-meaning parent-educators and professionals imply that kids throw tantrums intentionally, they're revealing a fundamental misunderstanding of how little humans develop control over their impulses, emotions, and behaviors. My aim in this book is to provide context through an understanding of the human nervous system—to help you as a parent understand how children develop self-control and emotional flexibility, and what you can do to foster that growth, given your child's unique makeup and basic genetic blueprint.

The most popular thing I ever posted on social media was a single sentence, something I often say to parents of frustrated toddlers: *"If the ability to control emotions and behaviors isn't fully developed until early adulthood, why are we requiring preschoolers to do this and then punishing them when they can't?"* Those words, displayed over a child's crayon drawing of the sun, have been viewed by more than two million people. Why did that statement resonate? Perhaps because its simple truth removes some of the self-blame and self-criticism that plague so many parents. (In fact, the ability to control our emotions and behaviors isn't a developmental milestone but an ongoing process informed by the body and brain communicating with each other, keeping us safe and constantly predicting what's going on.)

Most of us are never taught about the *context* of children's behaviors; we're taught about what to do to *manage* the behaviors. But there's a bigger picture involved: the *reasons* humans do what we do. Think of behaviors as the tip of an iceberg—the 10 percent or so that's visible above the waterline. Of course, there's much more to the iceberg. Hidden below the surface is a much larger piece, concealed from view but far more significant. When we react only to the behavior we see, we're overlooking this hidden part, ignoring the valuable information that can help us understand the "why" of the behavior, the rich clues about what is triggering such behavior. As a culture, we are extremely judgmental about behavior, especially our children's. And now, thanks to discoveries in neuroscience, there's a new and exciting story to tell about behaviors, feelings, and emotions and what drives them.

No matter what the behavior, there's much more happening than meets the eye. Our brain and body are constantly talking to each other—brains don't exist on their own! Children rarely act out for no reason or simply to make their parents' lives more challenging (as much as it sometimes feels that way). Our children's behaviors are outward signals of their internal world, indications about that submerged part of the iceberg. We should value behaviors for what they tell us about the child's body and brain. *Instead of trying to eliminate the behaviors, we should strive to understand them for the rich information they offer about how our child experiences the world.*

No matter how much you try to reason, reward, or offer incentives, you can't coerce or even teach a child to have control over something they can't actually control. What *can* you do? Instead of trying to correct or eliminate a concerning behavior, try to understand the clues it offers about your child's inner experiences.

THE PLATFORM

In particular, behaviors provide clues about the state of a child's autonomic nervous system, the unique two-way system of communication between the body and brain. *The brain-body connection, our nervous system, serves as a neural platform that influences human behaviors.* The child's body and brain are linked in a constant feedback loop through this system. So it's incorrect to consider the child's thinking or emotional expressions separately from the state of the child's body. The state of the body influences the way we feel, act, and think. We will refer to this complex and extraordinary system as the **platform**.

Because we are never just a "body" or a "brain"; we are always both.

Each of us reacts to the world from moment to moment on a continuum from receptive to defensive. When we experience a challenge as a fear or threat, we're in a defensive mode. When we're feeling safe, we're in a receptive mode. In my extensive work with children, I have witnessed that what influences a child's level of receptivity is the state of their autonomic nervous system, the platform. A *sturdy* platform supports optimal behaviors and strengthens the child's capacity to be flexible, think, and make decisions. A *vulnerable* platform, on the other hand, increases a child's wariness, fear, and defensiveness. When a child's platform is vulnerable, we see the behaviors that confuse and challenge parents: refusing to wear socks or to eat any food that's green, smacking a sibling, or throwing the remote when it's time to turn off the TV. These behaviors appear to reflect a child being oppositional, uncooperative, or impolite. On the opposite extreme are those moments when our children check out and disconnect, seeming to ignore us. The point here that I will unpack in the following chapters is that behaviors that appear to be defensive can actually be protective.

Another example of vulnerability in a child is hypervigilance, which can show up as overcompliance, a possible signal that the child is too concerned about pleasing others. While this behavior is often rewarded, it can represent a vulnerable platform. Children with vulnerable platforms are prone to being vigilant, worried, and disagreeable as well as yelling, crying, having a tantrum, running away, striking out, or even shutting down. Humans are not always in intentional control of our behaviors—children aren't necessarily consciously *choosing* these behaviors. Rather, many reactions and behaviors serve to protect the child from a deeply and subconsciously felt sense of unease or threat.

Here's a remarkable insight: We can understand a child's level of sturdiness or vulnerability by tracking what's called *allostasis*, the process by which we maintain stability in our bodies. But you don't have to remember that scientific word! The neuroscientist and researcher Lisa Feldman Barrett has another word for this continuous balancing of energy and resources: *body budgeting.* Just as a financial budget keeps track of money, she says, bodies track "resources like water, salt, and glucose as you gain and lose them." Although we are not always aware of our body's metabolic budget, *everything we experience*, including our feelings and actions, becomes deposits or withdrawals in our *body budget*. A hug, a good night's sleep, playing with friends, and a healthy meal: All of these are deposits. Then there are withdrawals: things like forgetting to eat meals or drink enough fluids, being deprived of deep sleep, or being isolated or ignored.

In the coming pages and throughout the book, we'll use Dr. Barrett's useful term, the *body budget*, to help guide us in making big and small parenting decisions. We'll customize those choices based on our child's body budget—and our own.

Parents are constantly faced with dilemmas. When a child faces a particular problem, should we encourage the child to work it out on their own, or is it more appropriate for us to make a "deposit" into the child's budget with our supportive and loving parental interactions?

The child's platform reflects the child's body budget and helps us make these decisions. In the next few chapters, we'll learn how to track our child's behaviors and other signals, helping us to discover what resources are available to the child in the moment and cumulatively. We do this by getting to know what our child's body is telling us through their verbal and, most important, *nonverbal signals.*

This is the central idea I'm excited to share: Our best parenting decisions aren't focused simply on our child's behaviors or thoughts but rather on our *child's body and the unique way each child continually processes, interprets, and experiences their world.*

That's why our strategy as parents shouldn't start with trying to *eliminate* a given behavior. Instead, we should work to strengthen our child's platform (and our own). We should start with these essential questions about how we customize our approach in any given moment: What is this behavior communicating about what my child needs from me right now? Does she need my words and my talking to her? Does he need a hug or a shoulder to cry on? Do they need me to set limits and remind them of consequences? Or does she need something more basic to get to a strong platform? Does he need a "top-down" thinking approach—reasoning with the child—or a more "bottom-up" body one, which entails strengthening the platform first? Or perhaps a hybrid of the two approaches?

Each child, each situation is unique. Most parenting programs fail to raise these questions, which are essential. For a child to make use of the information we want to impart, the child needs to have a sturdy platform. And we don't build a strong platform by incentivizing, ignoring certain behaviors, punishing, shaming, or talking at children. We do so by being present with the child and nurturing the relational trust that comes from our loving and consistent presence, customized and *personalized to our child's individual needs*, while keeping in mind that our job as parents is to help our child grow increasingly more flexible and resilient. With our presence as the cornerstone, then we can extend what we ask of the child and help the child stretch and tolerate new experiences as they grow more and more self-sufficient.

Here is a new way of thinking about common parenting challenges: What we view as behavioral or emotional challenges often represent an adaptive, subconscious response to the child's inner reality. It's the way the child's body is reacting to changes that require an adjustment or response—in other words, *to stress.* If we think of stress as our body's natural adjustment to changes, we can appreciate behavioral challenges for what they tell us about how our child's body and brain are managing what we ask of them. *What we'll typically find is a depleted body budget driving the "bad" behaviors, which aren't actually bad but subconsciously protective.*

That reality is often hidden from the adults in the child's life—masked by our own annoyance at the behavior—until the adults begin to ask themselves *why the child is doing what the child is doing, and how that particular behavior helps the child cope.* In this new way of thinking, we see children's behaviors as evidence of the power of human adaptation. It's helpful to move away from categorizing behaviors as either "good" or "bad." They are *adaptive*—and, for parents, sources of incredibly useful information.

Behaviors offer valuable clues about the state of the child's platform and what the child needs from us. When we shift our thinking, we can see that we miss the mark when we focus on eliminating a behavior instead of asking what it can reveal about the platform. Instead of telling a whining child simply to stop, we might consider that the whine is a signal that the child needs extra reassurance to feel calm. Telling a child to sit still at the dinner table doesn't help when she feels the need to move her body in order to manage her stress. If a child feels afraid of something that seems to be unthreatening, such as going to soccer practice or the sound of a particular toy, it's not useful simply to say, "There's nothing to be afraid of." We need to listen to what that fear is revealing to us. *The behavior offers clues as to what's going on inside the child, so this is an essential moment to pause to reflect about what's happening on a deeper level.* Once we understand how our child's body and brain are managing life's big and little challenges, we can help children to use new experiences to grow and not be overwhelmed by them.

The Platform as a Road Map to the Just-Right Challenge

We want to see our children thrive with each new skill they acquire, whether it's successfully nursing, swallowing their first bites of food, their first steps, the first day of school, or their first experience at summer camp. We help children tolerate these new experiences and situations by striking a balance. *We need to make sure that they are challenged enough to grow new strengths, but not overwhelmed by what we are asking of them.* To do this, we need to find what has been called the *just-right challenge*. I call it simply the *challenge zone*, where our children gain new strengths (because that's where growth happens), learn new things, and reach their potential with the proper level of support. How to detect that zone? We follow the many signals a child's body communicates.

A child with a depleted body budget is usually working outside of the challenge zone. It's important to define this zone for each child, because children can't grow to be resilient if their platforms are constantly overwhelmed *or* if they have too much support from adults constantly hovering over them, protecting them from healthy challenges. Throughout this book, I'll provide ideas and examples for how you determine the best challenge zone for your child, depending on the circumstances. In doing so, I'll address questions such as:

- *What do I do when my toddler (or child of any age) has tantrums?*
- *How do we manage sibling rivalry?*
- *How can we help our child sleep through the night?*
- *What do we do when our child is noncompliant or defiant?*
- *What do we do when our child faces problems with teachers or peers?*
- *How do we know if a limit we set is too firm or not firm enough?*

When we discover a child's challenge zone, we can gain the answers to these and other questions, and find confidence in parenting decisions that are personalized to each child's platform. And working within a child's challenge zone builds *frustration tolerance*, a child's ability to work with their frustration rather than giving up or melting down. A child with frustration tolerance can delay gratification and wait for what she wants, remaining calm when encountering obstacles.

As a parent, you play a crucial role in building and strengthening your child's platform. That reality doesn't have to add to the stress you already experience as a parent. As we'll see, it's never too late to strengthen the platform. Every interaction you have with your child can build the child's receptivity and resilience. *The window to helping our children thrive is always open.* And we don't have to be perfect about it. Fortunately, there's plenty of room for error; it's a learning process. What matters isn't that we do everything perfectly as parents. That's impossible! Rather, we should recognize when we've missed the mark, repair the situation, and learn from it. When we do, we and our children grow stronger as a result.

The more we understand how interconnected the body is with the brain, the more we realize that the state of the child's nervous system should be the primary factor informing our moment-to-moment parenting decisions. Any approach to parenting needs to consider three crucial factors: (1) the state of your child's and your platform (from sturdy to vulnerable), (2) your child's developmental abilities, and (3) your child's unique qualities—the *individual differences* that affect how children process information through their senses and from inside their bodies. Throughout this book we will use these three concepts to explore ways to customize your parenting to the particular needs of your child—as Leanda and Ross learned when their daughter Jade showed such resistance to kindergarten.

How Understanding the Platform Helped One Child

When Jade repeatedly begged her father not to leave her at kindergarten, both parents initially assumed that the daily struggle reflected a conscious, willful effort on Jade's part to avoid school. I wasn't so sure. After speaking to Jade's teacher, I discovered that most days she acted sad and quiet for about an hour, but after that she usually joined a friend in the classroom's play kitchen. Giggling with her pal and making pretend food for her classmates, Jade was a "different child," her teacher reported.

That was what I saw when I spent some time observing her in the classroom. Clearly, Jade loved aspects of school, but it took a toll on her. When she gyrated her body, tried to run away, clung to her dad, or screamed, it indicated that the drop-off caused a heavy cost to her body budget. Most of the time, no amount of cajoling or incentives (like putting stickers on charts or offering another form of reward) alleviates such situations. What *can* help is learning about how a child's platform functions depending on the situation.

After several long sessions of observing Jade and listening to her parents, I suggested that they view her actions not as reflecting a desire to avoid school but rather as *signs of her nervous system valiantly trying to manage the stress her body was experiencing.* Our task, I explained, wouldn't be simply to change the behavior but to help Jade at the level at which the behavior was generated, and, in doing so, to strengthen her brain-body platform—to help her move from a defensive, vulnerable place to a receptive and sturdy one. When the behaviors stopped on their own, that would be the sign that we had helped her platform.

We met with Jade's teacher and devised a plan personalized to Jade's experiences. The first step was for Jade's teachers to stop forcibly prying her from her father every morning. Instead, we needed to transform the routine from a "cold" hand-off to a "warm" one. Rather than leaving Jade in the chaos of the drop-off circle just when throngs of other students were arriving, Ross began delivering her fifteen minutes earlier to a quiet spot just outside the classroom. There the teacher could kneel and greet Jade with her warm, prosodic voice. The teacher chatted with the dad, giving Jade's platform a few minutes to warm up, and, finally, the teacher let Jade indicate when *she* was ready to say goodbye to her father and enter the classroom to help prepare for the day. The teacher took her cue from the student, respecting and honoring the girl's platform. It was only a week or so before Jade stopped begging her dad not to leave her, an indication that she was ready to go in on her own.

Just three weeks into the new plan—which included *no* mention of incentives or consequences—Jade volunteered to her father that she was ready to transition back to the regular drop-off circle. What had made such a difference? Instead of simply offering her rewards for not getting upset, we recognized that we first needed to *strengthen her platform*.

Understanding the importance of strengthening their daughter's platform helped Jade's parents know what to do in other similar situations. Jade loved to dance, yet when it came to trying a dance class at a local community center, she froze entering the room, begging her mother to join her. Wanting to be supportive, her mother tried the popular techniques for helping children deal with fears: encouraging her to name the emotion, discussing fear and how to combat it with calming thoughts, reminding Jade that Mommy was just outside the door. To be sure, these are viable tools that are effective once a child's physiology is calm enough to connect the body to the brain, but as with the kindergarten drop-off, Jade simply wasn't there yet.

The problem? Using a top-down approach for a body-up problem too soon. The solution? Focus on the platform! Since parents weren't allowed in the dance studio and had to wait in a nearby room, Jade's mother decided to let her teacher know about what worked at school. The dance teacher met with Jade just before class and gave her a job as a special helper. A few minutes of connection (perceived as a big deposit into her body budget) was all it took, and Jade accompanied the teacher into the class without her mother, holding the teacher's hand. This simple, human-to-human contact helped boost Jade's platform and enabled her to participate and enjoy the experience.

Of course, each child is different, but understanding a child's nervous system helps reveal how a child will receive our interactions and what we are asking of them. And the ability to read a child's need for emotional support as early as possible eventually leads to the child being able to participate in friendships, excel in school, and eventually become independent and resilient, able to manage life's struggles. It takes time and is influenced by how the relationships in a child's life nurture the child's biology. When we consider the child's platform first, before we ask the child to comply with what we are asking of them, we open new horizons in parenting.

HANG ON! IS THAT CODDLING?

Parents who are accustomed to more traditional parenting approaches may wonder whether the brain-body approach might be a form of coddling or permissive parenting. Personalize your parenting? Does that mean setting low expectations or paving the way and making things too easy for the child? Shouldn't parents have certain expectations?

Those are great questions. It's important to understand that the goal *isn't* to pave the way for your child to act out without consequence or to avoid ever stepping out of the comfort zone. In fact, our aim is quite the opposite. Most parents want to support their child's future resilience and independence. When you let the child's nervous system serve as a road map, you can develop a better sense of when it's appropriate to pull back and soothe the child, change plans to make the child feel safer, or let the child struggle through a challenge. Human beings don't develop new strengths without experiencing some degree of stretching or even discomfort. *The key is that the support needs to be tailored to the child and to the circumstances.* Otherwise—as Jade's experience shows—many approaches based on simply trying to alter a child's behaviors fall short or even backfire.

PARENTS HAVE PLATFORMS, TOO

Needless to say, we all have nervous systems—adults, too. Our children's platforms influence us, and ours affect them. As we will discover, our own platforms matter when it comes to parenting. (Chapter 5 will describe how we can nurture ourselves and make deposits into our own body budgets to balance out the constant withdrawals that come from parenting.) Many of us may not have had parents who nurtured our nervous systems by attending to our emotional needs, instead labeling or judging our behaviors and emotions.

Many of us recall our own parents saying things like, "Don't cry—it's not so bad," or "Lots of kids are worse off than you," or "Come on, there's nothing to be afraid of!" The well-intentioned message had an unintended consequence: to ignore our bodies' signals and sensations of distress, not even considering the feeling states underlying our behaviors. Even now, our parenting and education cultures generally don't appreciate the deeper meaning of "negative" emotions and behaviors, and what they can tell us about the stress a person is experiencing in their body. Over time, ignoring and undervaluing these expected human responses to stress can lead to various health conditions caused by *allostatic load*, or the effects of long-term stress. These can include inflammation, high blood pressure, heart disease, eating disorders, anxiety, or depression. Earlier generations did their best with the information they had. Now, though, we know more about how stress manifests in the human body. *When we consider the body and brain in a more integrated way, our children's and our health and well-being benefit.*

I have many memories of painful parenting moments that resulted from my own sometimes-weak platform. One in particular stands out in my mind. I was picking up my children from school after a particularly stressful day at my office. What I hadn't realized was that the stress had degraded my own platform. When I arrived, my four-year-old decided that she didn't want to get into the car, instead begging to stay and keep playing with a friend. After a few minutes of conversation and debate about why we needed to leave *now*, my platform broke. Suddenly I lost control over my emotions, grabbed my daughter, and yelled at her at the top of my voice. At that moment, I looked up to see a mom I knew staring at me in surprise.

(I still remember the look on her face all these years later!) I sat in the car, feeling a pain in my gut. What had I done? My children, now all in the back seat, were quiet. I had scared them, and I felt full of regret and, most of all, shame. *What kind of a mother,* I thought, *a psychologist, no less, scares her own children?*

As the evening wore on, I apologized and tried to get the girls to talk about their feelings and memory of the event. They were understanding but didn't want to discuss it. The next day, I visited an art studio near my office where I sometimes dropped in to paint designs on ceramic pieces—an activity I found relaxing. That day, I decorated a square coaster with three words, "Handle with Care," and my three daughters' names. As I sat in the studio, I promised myself never to "lose it" again in that way with my children. All these years later, I still keep that coaster on my nightstand, a visual reminder of my vulnerability and my intention. This memento still reminds me to treat my children, and myself, with love and compassion.

Indeed, those two powerful forces—love and compassion—are at the center of this approach to parenting. *Shifting from addressing behaviors to trying to understand their origins and triggers means making a shift from managing our children to understanding them deeply.* We can open ourselves to parenting with a sense of reflection and curiosity, observing ourselves and our children in a new way. *We can move from putting out fires and reacting to situations to pausing and questioning what our children's behaviors are telling us about their bodies and brains.*

THE GROUNDBREAKING SHIFT IN PARENTING FOCUS: FROM BEHAVIORS TO NERVOUS SYSTEMS

To summarize, we need to shift from focusing on children's behaviors to parenting their nervous systems—and caring for our own. We need to nurture our children's physiology and psychology (platforms) by paying attention to their body budgets and what's causing the behaviors we observe. *As a psychologist, I no longer work on behaviors. I work on supporting nervous systems and looking to the underlying sources of challenges.*

When we appreciate how humans develop, from the body up, a whole new horizon of positive parenting tools based on an appreciation for the mind and body becomes available to us. We build strong platforms through the way we *nurture our child's nature* and as we endeavor to match our parenting with the shifting needs of the child we are blessed to parent. In turn, we'll learn to find personalized answers to our parenting challenges.

The most important factor in strengthening a child's neural platform is understanding the mechanics behind how humans learn to feel safe in their bodies and in the world. To understand that, we'll learn in Chapter 2 how our nervous systems detect whether a situation is safe or dangerous, comfortable or life threatening—a phenomenon called neuroception.

RESILIENCE-BUILDING TIP: Children's (and our own) behaviors are an outward reflection of the complex workings of the brain-body connection, their platform. When we stop to consider what their behaviors are telling us about their platform, we have our first clue for building resilience.

2

Neuroception and the Quest to Feel Safe and Loved

> Before they can make any kind of lasting change at all in their behavior, they need to feel safe and loved.
>
> —Dr. Bruce Perry

Lester and Heather had just moved across the country when their eight-year-old son started to have difficulty sleeping. Randy was excited to have his own bedroom after sharing a room with his toddler sister in their old house. He had even chosen the paint colors and selected a superhero-themed comforter for his bed. But soon after the move, Randy started waking in the middle of the night in distress and appearing at his parents' bedroom door.

He also became inexplicably obsessed with keeping the floors clean, vacuuming the carpet when his sister spilled crumbs—or for no reason at all. At first the parents were amused by his new cleaning habits, but then when he added a new behavior—spending long hours ritually organizing his sister's toys—they contacted me for help.

His parents had tried discussing the matter with Randy, encouraging him to express his feelings and reasoning with him about his late-night fears. They installed a night-light and tried enticing him to stay in bed by offering to reward him with outings to a frozen-yogurt shop. None of that helped. In just a few months, their previously happy child had turned clingy and seemingly regressed in maturity.

Sometimes, our usual parenting tools—reasoning, limit-setting, encouraging—fall short. That's often because what's driving the child's behaviors is a deeper need beyond the child's conscious awareness: the basic human drive to feel safe in the world. Indeed, from the moment we are born, humans are on a quest to feel safe, a quest hardwired into our basic physiology.

As we have seen, when a child has a strong platform, they are more likely to be cooperative and able to respond to a wide range of requests. But if a child's platform is vulnerable, the range of what they are able to manage narrows, sometimes so much that they might feel overwhelmed by even a simple task. That explains what happened to Randy. After the family's move, he seemed more dependent on his parents and less able to enjoy things like his own bedroom or the chance to meet new friends or explore his new neighborhood.

I reassured Lester and Heather that there were good reasons for what they had observed in their son and that we would work together to find answers. Each child's platform is determined by the child's unique experience of the world, and it's not uncommon to experience challenges, especially when we face significant changes in our lives. These temporary setbacks offer the chance to better understand a child and discover what can make them feel safe. Eventually, that knowledge will enable us to help the child adjust to life's demands.

Feeling safe allows children to work at the top of their emerging *executive functions*. (Think of those things successful executives need to do to run businesses: stay focused, meet challenges, have self-discipline, and flexibly adapt to changing circumstances.) Of course, we want our children to develop these essential skills for future independence and success. But it's important to remember that children build them over years—even decades. Throughout childhood and into early adulthood, children develop and strengthen the ability to control their emotions and behaviors, think and plan, and adapt to changes and circumstances. It's not unusual for challenges such as family relocations to present temporary setbacks. As I explained to Lester and Heather, their son's behaviors would likely make sense to them once they considered how children adapt to challenges and to stress.

As we learned in Chapter 1, the brain constantly monitors our body budget to make sure we are in balance—or, in scientific terms, maintaining allostasis. We don't have to will our heart to beat, our lungs to breathe, or our digestive tract to process food. Our nervous system makes this happen by constantly monitoring information coming in from the environment and from our internal organs, making adjustments as needed to keep us healthy and safe.

Our bodies are equipped with a remarkable monitoring system that determines whether we will be receptive or defensive, cooperative or uncooperative. Understanding this can help us personalize our approach to our child's reactions. Ignoring it can lead us to demand more of children than they can offer.

To begin, we need to understand how humans take in and understand our world. Once we understand how a child is reacting to a situation, we can try to bolster the platform (give them a hug, slow down) or move ahead with our request or expectation and encourage the child to take on a challenge more independently. It helps us to determine: Am I asking too much of my child in this moment? To understand how to answer that question, we need to understand the brain and body's threat-detection and safety-detection system.

THE DETECTION OF SAFETY

Neuroception, Interoception, and the Safety-Detection System

Deep within our nervous system, out of our awareness, every human being has an innate process that detects safety or threat. This hidden sense kept our ancestors—those who survived—alive over many thousands of years, allowing them, for example, to flee a wild animal or an oncoming fire instinctively, escaping quickly and efficiently.

Dr. Porges coined a name for this ability to detect threat and safety: neuroception. It describes the way our nervous system constantly surveys our external environment (outside of our bodies), our internal environment (inside our bodies), and our relationships with other people to make sure we are safe, and it instinctively directs the body to action when we are not. Neuroception describes how our brain automatically and subconsciously makes sense of sensations and judges them as safe or not safe. Although the sensory signals triggering neuroception are outside our awareness, we are frequently aware of the *impact* of neuroception as feelings in our body (e.g., increases in heart rate or heart pounding). These feelings arise through another process called *interoception*, through which signals from deep within our bodies can alert us to how we feel on the inside. This is called interoceptive awareness. Complicated words, right? If you've never heard of neuroception and interoception, that's okay. Let me explain and simplify what this all means because it's going to help you understand your child's (and your own) reactions to life's experiences in a new way.

How does neuroception work? The neuroscientist Lisa Feldman Barrett explains that our brain is basically a lightning-fast prediction machine. Outside of our awareness, different parts of the brain talk to each other, constantly evaluating the flood of incoming sensations against all past experiences to instruct the body what to do next. "Your brain is always predicting, and its most important mission is predicting your body's energy needs, so you can stay alive and well," she writes. This is what happens when you hear the sound of a car that's coming too close and, before you know it, your body has jumped out of the way. You didn't consciously plan the action; your brain did it for you to protect you. That big jump was a cost to your body budget, but it might have saved your life.

Understanding how our brains make sense of the information coming from inside and outside of our bodies is crucial to understanding our children's emotions and behaviors, as well as our own, helping us perceive a child's relative receptivity and accessibility. It provides a neuroscientific basis for the idea that safety and trust are central to human development. *And it helps parents understand how a child is perceiving their world and how to tailor our interactions based on that information.*

Think of neuroception as a computer program that's constantly running in the background, designed to spur us into action (or inaction) to keep us safe. I'll use the terms *safety-detection system* and *safety system* interchangeably with the more scientific term *neuroception*. Our safety-detection system's full-time job is to determine if the sensations we experience from inside our bodies, in the environment, or from other people represent safety or threat, and, if it detects threat, it tells the body what to do quickly and most efficiently.

How? We survey and interpret the world through our *sensory systems*—what we feel, see, hear, smell, taste, and touch, among other senses. We take in the world, including internal sensations (such as aches or hunger pangs through interoception), our immediate environment (such as noises or aromas), and cues from other people (such as how they look or talk to us), under the surveillance of our safety-detection system—our neuroception. In Chapter 6, we'll cover in detail how your child takes in sensory information, because understanding that is key to understanding their emotions and behaviors. For now, though, what's important to know is that what *you* perceive as a challenge may not be what your *child* perceives as a challenge or threat. *That's because of individual differences, the impact of our past experiences, our genetics, our constitution, and the wide variety of human experience.*

Neuroception is unique to each person. We each experience internal and external sensations through our own individual brain-body feedback system. Some people pop a Tylenol at the first sign of a headache, immediately feeling anxious about whether it will worsen or improve. Others might not be bothered by the same degree of headache. One person might experience a particular circumstance or sensation as intense or painful while another might tolerate it easily.

This goes for our children as well, which is why one child might resist viewing a particular kind of movie (because their safety-detection system interprets loud sounds as unpleasant or scary and threatening) while another begs to watch the same film (because their safety system judges the sounds as enjoyable and safe). In other words, it's the *detection* of safety and not the objective circumstance that determines whether or not we feel safe and what behaviors result from our body's subconscious surveillance system.

Parents often tell me that their child's troubling behavior seems to occur "out of the blue." The concept of neuroception, though, shows how inaccurate that is. Children are always reacting to something, even if that something is invisible to us as parents—and to them as well. Neuroception explains the "why" behind a child having a negative or positive reaction to life's events, big and small, and shows that behaviors rarely happen "out of the blue."

For example, your toddler hears the sound of a helicopter from afar that registers as unpleasant in their body and starts to cry (neuroception of threat). You don't notice it so you are confused as to why they're crying. Then you finally hear the helicopter and realize they're troubled by the sound. You comfort your toddler and then, minutes later, their body picks up all of your cues of safety, and they're running around and smiling again, recovered (neuroception of safety).

Understanding neuroception is useful in answering many in-the-moment parenting questions. When we ask an essential question, *"Is the child perceiving safety in their nervous system?"* the answer yields a road map to use in that moment. Often, a child can't actually answer the question with words, but we can see the signs of distress in the child's body—such as the way the toddler reacted to the helicopter sounds. We start by considering what cues of threat the child might be experiencing, their sources, and why the child feels vulnerable.

I reassured Heather and Lester that their son Randy's seemingly odd behaviors—not sleeping independently and obsessively cleaning—were a signal that his safety system was doing its job, detecting threat and then directing him to react in a way that helped him feel safer. Our job was to help him turn his subconsciously driven sense of unease into *perception*, something he could manage and talk about through awareness. This way, we could help Randy expand his tool kit for managing the stress, giving him a way to help himself feel better in addition to cleaning and organizing. *Most important, we could help him develop an appreciation of his body's need for support and learn how to verbalize his discomfort and connect with his parents more directly during stressful times, a most helpful tool for all of us to have.*

Each of us has our own detection of safety and threat because each of us has a unique experience of the world from birth—and even in utero. One of my children, for example, was born prematurely and had hair-trigger reactions to the world that tilted toward picking up threat from her environment early on. Her systems simply weren't quite ready for the lights, sounds, smells, and movement she encountered outside the womb. So when she was a newborn, her safety system detected threat from such ordinary experiences as a light being switched on or even my energetic manner of speaking to her. She was easily rattled because her brain and body weren't ready to handle the incoming stimulation from the environment or, sometimes, from her eager and doting parents. Her body had difficulty adjusting to experiences, and only later did my husband and I discover that we had been inadvertently setting her off with our exuberance and by offering too much movement and too many sounds and types of physical stimulation too quickly. *We thought we were making deposits into her system, but we were inadvertently making withdrawals!* We had no idea that often, well-intended interactions sent her nervous system into distress.

My daughter had *sensory over-reactivity*—or big reactions to everyday sensations. Her safety system would detect threat even when she was objectively safe. The results were exaggerated reactions to seemingly innocuous situations, such as when we'd move her or sing or talk too loudly, too quickly, or too close to her. At the time, years before I integrated the *body's reactions and needs* into my practice of psychology, I didn't yet understand that the sound of my own voice could trigger my tiny baby's body into distress.

Why would a child's nervous system erroneously detect threat when the child is actually physically safe? For a variety of reasons. Genetic or constitutional factors can predispose a child to sensitivities that make the child overreact to sensory experiences. Past experiences influence children and help them predict what will happen in future similar circumstances. That was the case for my premature baby. Her reactions were based on her genetics along with her earliest subconscious memories of medical procedures, which played a role in how she interpreted sensations and how her body tried to protect her—a tribute to the body's survival instincts. Sometimes early experiences, such as invasive or painful medical procedures, excessive stress in the child's environment, or losses or separations, can cause a child's brain to predict danger or threat more often, causing stress to the child's system. *And sometimes, it's simply the way your child is taking in information and reacting to it.*

As we'll cover in more detail in Chapter 6, every human has unique reactions to sensory and other information that's coming from *inside* the body, such as a rapid heart rate or hunger pangs, as well as to circumstances or changes that occur *outside* the body, such as living near a busy freeway, moving to a new house, or losing a loved one.

We're all familiar with toddlers crying when they enter a pediatrician's office, triggered by the memory of an immunization needle. Our early experiences form subconscious body memories that can be prompted by features in the environment that remind the nervous system of something the safety-detection system initially coded as threatening. We can't always shelter children from many of these experiences, but we can use their reactions as clues to help us support them emotionally.

Of course, the safety-detection system works the same way in adults. When my children were young, and I was extra tired and stretched, I could easily lose my temper over simple things. My body budget balance was always low because of the constant withdrawals from parenting three children and working as a psychologist. If we were late getting out the door in the morning, or in a public place where I was worried one of my toddlers might wander, I'd bark orders at them. When my body budget was in a deficit, I'd sometimes say things I later regretted, projecting my own lack of internal resources onto my kids: "Hurry up! You're making us late!" (Message:

stop being a slacker.) My depleted state was adding to my shame for being late, my own issue, and I was taking it out on my children. Lack of sleep, work stress, multitasking, and managing their lives and my own frequently turned me into an authoritarian and controlling mom, markedly different from how I was with them when my nervous system felt safe and secure. (Back then I didn't connect the words *nervous system* and *parenting* in a meaningful way.)

I often felt badly about my parenting and, after the heat of the moment, wondered why I had melted down. It wasn't because I didn't adore my kids. It was because I didn't know the toll my own busy life was taking on me. I had never heard of allostasis or a body budget, but the proof was right there: I usually lost my temper with my kids when I was running on a deficit of sleep, nutrition, or solitary time to recover and recharge my batteries.

THE SAFETY-DETECTION SYSTEM READING DETERMINES THE PLATFORM STRENGTH

The ongoing reading of our internal and external environments directs our bodies to engage in actions to feel safe again. If a child's safety system detects threat, we are likely to see behaviors resulting from that reading. The child might refuse or resist doing something, or have difficulty adapting to a situation, as Randy did after their family move. Tasks that he had previously mastered, such as sleeping independently, were suddenly coded as unpleasant or threatening. This is when parents can feel like they are constantly "walking on eggshells" around their child's reactions. When a child's nervous system struggles to maintain a body budget, that's when we witness what we may interpret as "negative" behaviors: being uncooperative, being overly controlling, fussing, complaining. They can escalate to more serious reactions: yelling, hitting, running away, or throwing things.

It's probably becoming clear that stress can weaken the platform. But not all stress is bad. In fact, without stress, development stalls, because learning and growth require experiencing changes. Recall from the previous chapter that stress involves the body and brain reacting to life's challenges and changes. We need a certain amount of stress for the brain to realize that there's something new to pay attention to. That's how we learn new things.

Stress that is *predictable, moderate, and controlled* leads to resilience. Think of a kindergarten classroom, where children's mild stress from being away from their parents mixes with the novelty of learning new facts and making friends. This is within most children's challenge zones, and throughout their educational experiences, they keep learning because of this tolerable, moderate, and predictable "good" stress. Being able to stretch over time and not become overwhelmed is what builds resilience.

Tolerable and predictable stress is what helps children grow and develop new strengths. But when stress is *unpredictable, severe, and prolonged*, it threatens resilience, and children or adults begin to suffer the consequences of the chronic activation of the stress response.

If we take good care of our bodies and minds, we support our physical and mental health. But if your stress—or your child's—arises too frequently, is too intense, or continues for too long, the body budget can run a deficit. When stress adds up day after day, year after year, the cumulative impact—known as the allostatic load—can be damaging. The physical toll adds up over time and can contribute to diseases with heavy links to stress, including high blood pressure, heart disease, obesity, type 2 diabetes, depression, anxiety, and other conditions that are prevalent in older adults. In subsequent chapters, we'll see the many ways parents make deposits into our children's—and our own—body budgets so that they run a good balance. But the most important way for all humans to maintain their body budgets is to get enough sleep. Sleep sets the foundation for successfully managing life on all levels, and for all ages.

Children who experience prolonged and extreme behavioral or emotional challenges may need help in managing their stress load. It's rarely a single incident that shifts a child into distress. Usually it's a combination of factors beneath the tip of the iceberg—the child's experiences in recent days, inadequate sleep, how the child feels physically, and whether the child has expended energy on stresses beyond your awareness. *The behaviors we see represent a child's, or our own, accumulating stress load.*

The safety-detection system helps us understand why the same experience can send your child into a meltdown one day but seem manageable the next. What's different is where it's landing in your child and the balance in their body budget based on the *accumulation* of experiences that your child continually takes in as either safe or threatening.

The detection of threat and safety is unique to each of us. The same experience one person finds stressful or threatening may register as safe in another person. We each respond to the world in our own way, depending on *countless* factors including how we take in information from the world through our senses, all of our past experiences, and the state of our body budget. That's why it's so important to personalize our parenting to our child's unique reactions. To illustrate how unique the detection of safety is for parents and children alike, let's examine how two families dealt with challenging situations in very different ways, and then we will return to Randy's story and how his family supported him after their cross-country move.

TWO CHILDREN AND TWO FAMILIES WITH DIFFERENT EXPERIENCES

Parker

In the first months of Parker's life, his doctors discovered he had a heart condition that would later require corrective surgery. When he was three, his pediatrician suggested that his parents meet with me to discuss how to emotionally prepare Parker—and themselves—for the operation. At our first meeting, without their son present, his parents described how the diagnosis had rocked their world. But they also described their strong family and a faith tradition that gave them conviction that a higher power was guiding Parker's medical journey.

While the situation would have caused stress or even trauma for many parents, Parker's parents were not registering the experience as such. To be sure, they had experienced fear and shock, but they also benefited from many layers of support—from family, friends, doctors, their faith, and each other. They seemed to have found a way to fortify themselves—and their child—for the surgery and recovery period. In home videos of Parker they shared, I saw an exuberant little boy playing with his doting parents in their backyard. After that, we met just a few times to discuss how to talk to Parker about the surgery.

Three months after the successful procedure, his parents brought Parker for a visit. Happy, joyful, and connected, he asked his parents to explore the playrooms of my office with him and then, smiling, pulled something from his mother's bag. It was a colorful photo album, its cover embossed with the words PARKER'S HERO JOURNEY. He happily showed me photos of himself in the hospital wearing a gown of fabric with his favorite superheroes. Turning the pages, he was visibly proud as he displayed the photos and shared memories, engaging in conversation with his mother and me. Clearly, he remembered the surgery as an adventure, not a traumatic experience. His parents hadn't interpreted the situation as an unmanageable threat, so it wasn't surprising that neither did their son.

To be clear, Parker's parents weren't in denial. It was a stressful event, and the family didn't gloss over the gravity of it. Parker experienced many moments of fear and distress, but his parents were always there, helping him manage those strong feelings and providing cues of safety (smiles, hugs, comforting words) that helped him cope. That support allowed him to modulate his reactions and interpret the unfolding situation as *manageable.* The proof was in the child's carefree demeanor and the photo album he proudly held. His parents, sustained by a strong support system, experienced what might have been traumatic for many families as a tolerable and strength-building experience for their son and for themselves. I let them know that they checked off all the boxes for resilience.

Rana

Rana was another child facing a medical issue. When she was two, her mother, Greta, discovered a lump in Rana's groin while bathing her. The mother's safety system detected a threat, which sent her stomach lurching. Within seconds, Greta told me later, her mind went to the most ominous possible prognosis: cancer.

As it turned out, Rana had no such serious medical condition. Doctors diagnosed an inguinal hernia—easily correctable through outpatient surgery. Rana's pediatric surgeon referred the family to me because the doctor was concerned about the emotional distress that he observed in Greta during Rana's preoperative visit and from the report that the mother had fainted during Rana's routine blood draw. Even though the information from the medical professionals was reassuring, this mother's safety sensors registered her daughter's relatively mild medical condition as a serious threat.

A brief procedure corrected the hernia and Rana recovered quickly. I observed her several weeks later, on the playground of her day care center, where she played happily with the other children. Greta, on the other hand, continued to suffer, continually worrying that something else might happen to Rana. After reassuring Greta that Rana was doing fine, I suggested that she and I continue meeting together. Over several months, I tried to help Greta make sense of the intense emotions and fear that her daughter's situation had stirred in her. It took some digging, but I eventually learned that she had lost a beloved aunt to cancer as a child, and memories of that experience predisposed her to fear and vulnerability around medical issues. *Over time, she came to understand that her prior experiences tilted her nervous system toward detecting threat when it came to managing her emotions around Rana's health and well-being.*

As the experiences of these two families illustrate, the impact of stress on each person is determined by how our nervous system *interprets* life's events—and not necessarily the events themselves. If our sense of safety was dependent solely on the events themselves, then, logically, Parker's parents—and Parker—would have experienced more stress, and Rana's mother less. His surgery was much riskier. But the opposite was true—because a human being's sense of safety is influenced by our past and present experiences.

This is why it's best not to judge our children's reactions automatically but rather to be mindful of where experiences are landing, with compassion and without judgment for them—and for ourselves. When children have a strong negative reaction to an experience, we can help them regain their strong platform again by addressing the source of the threat they are experiencing.

TWO STEPS FOR ADDRESSING THE FELT SENSE OF SAFETY

Now that we have a new understanding of the importance of the detection of safety, I'll describe two steps to help your child feel safe when the child's behaviors reveal that their platform needs some safety boosting. The two steps are to (1) *resolve or reduce* the cues of detected threat that are triggering the child (if possible and appropriate to the situation), and (2) *bring in* cues of safety that work for the child to counterbalance the stress.

Sometimes, it's both possible and advisable to wait and see if a child can work on and resolve a problem on their own. But other times, it's beneficial to identify and address or help resolve a cue of threat for a child. Here are a few examples of successfully meeting a child's safety needs:

Your one-year-old suddenly begins crying during a walk in the stroller. It's the kind of cry that you know signals that she's distressed and not simply tired or fussy. You just fed her and changed her diaper, so you wonder, What's set off her safety sensor? Looking more closely, you see that she's turning her head in your direction, but the stroller's sunshade is blocking her view of you. She registered unease because she couldn't see you. Kneeling by the stroller and smiling, you reassure her in a gentle voice: "I'm right here, sweetie. Everything's okay!" Soon, she begins to smile. You pull back the shade and notice that she looks up at you every few minutes, smiling broadly.

ADDRESS THE THREAT: You pulled back the shade so she could see you, helping reduce the uncertainty. A primary way babies and toddlers feel safe is by looking toward their loved ones.

ADD CUES OF SAFETY: You smile at her, use your voice to soothe her, and then continue to talk to her every now and then in a comforting tone to increase the safety factor. These are the cues that help her body register safety again.

Your five-year-old begins to yell and complain within minutes of trying on his new school uniform—the first he's ever had. He says he hates it and doesn't want to go to school. Aware of his aversion to certain rough textures in clothing, you gently acknowledge that he seems uncomfortable and wait for his response. He says yes and bursts into tears. You ask him if he can think of any solutions, engaging him in solving the problem. He tells you that he wants to wear an old T-shirt to school instead. Knowing this isn't an option, you remember a box of used uniforms in your garage and ask him if he would like to try a softer, pre-worn uniform that you remember a fellow parent has shared with you. His body visibly relaxes as he tries on the secondhand garment and then eagerly asks you how many "sleeps" until the first day of school.

ADDRESS THE THREAT: Noticing his heartfelt reaction to the physical experience of the uniform, you take seriously his body's reactions, slow down the process, and let him be a part of the solution: a softer alternative.

ADD CUES OF SAFETY: Instead of leaping to judgment, you stay calm and validate the child's sudden negative experience in his body's reaction. *You realize this was a body-up reaction and not a conscious choice to be picky.* You add cues of safety through your empathic tone of voice, your facial expression, and your steady presence and by providing a reasonable option.

Your ten-year-old suddenly becomes reclusive and withdrawn. After a few days, she admits that some peers have begun to bully her at school. You praise her for letting you know and ask for suggestions to create a plan to help manage the problem. She asks if she can email the teacher or meet with the teacher and you to discuss solutions. She then asks to invite a few close friends over on the weekend and plan some fun activities with these trusted classmates.

ADDRESS THE THREAT: Sensing something was bothering her, you give your child the time and space to talk to you about it. Your accepting tone allows her to define the problem and come up with solutions to address the problem proactively with you.

ADD CUES OF SAFETY: You know that she has a core group of trusted friends and nothing helps humans feel safer than others who love them. Your willingness to host the friends over the weekend adds powerful cues of safety to her nervous system, as she anticipates the support of her friends.

PAYING ATTENTION, WITHOUT JUDGMENT, TO WHERE EXPERIENCES LAND

When we pay attention to where experiences land emotionally for our children, we help expand their emotional range and build their tolerance for stressful situations. We strengthen children's platforms by becoming attuned to their *unique interpretations* of life's events instead of defaulting to assumptions about what we *think* their reactions should be. We can then respond based on a compassionate understanding that it's often the child's nervous system, and not the child's will, that opts for seemingly negative reactions and behaviors. *We can see that a child who struggles isn't necessarily choosing to be difficult but is experiencing a stress response.* This adjustment helps us avoid making judgmental statements to ourselves or aloud (e.g., "They're overreacting," "Just get over it!," or "Tough it out!").

We must also be careful not to judge a child based on the way they express their emotions. The expressions that manifest on the outside may not offer an accurate mirror for the feelings on the inside. For example, you may see a child laugh at something serious, in a way that feels "inappropriate" to you. In the same way, a child with a frown could be angry, frustrated, or concentrating—or be experiencing some other emotion or sensation. *Our interpretations of their behavior assigns a meaning that may not be accurate, and our own reaction to their expression may prevent us from supporting their underlying needs.* This is why it is so critical to discern how their nervous system is perceiving the situation.

The key here is understanding the wide range of variability in how children respond adaptively to how their bodies are experiencing the world. Rather than judging their reactions as appropriate or inappropriate, we can view these trying behaviors as signals that our child needs relational deposits (an understanding look, a soft and caring voice, or a hug) and not withdrawals (a time-out, a lecture, or punishment). This runs counter to the way our culture typically views behaviors as either "good" or "bad," compliant or noncompliant. *It's a paradigm shift in how we view and judge behaviors.*

Each day brings opportunities to help build a child's sense of safety and trust in the world. One of the most powerful things we can do as parents is *acknowledge a child's emotions and automatic reactions as meaningful.* Many of us weren't raised this way. Our well-meaning parents may have said things like "There's nothing to be afraid of, sweetheart" if we had what they deemed as an irrational fear. What we can communicate instead to children is an appreciation of their distress and a validation of their safety. We might say, "I see you are struggling with this. I'm here with you. You're not alone"—simply acknowledging without judgment that we see they are struggling and that we are here to help.

Another way to create a sense of safety is to build predictability and flexibility into the fabric of your family's life. Nothing sets up a body and brain to feel safe like knowing what to expect. Humans love patterns and feel safe when life presents a pattern of predictability, in which our expectancies are met. We love soothing patterns because they take away uncertainty, which most people find unsettling. Think of the last time your plans suddenly changed and how your child reacted. If it was a negative reaction, it was likely because the pattern the child was expecting shifted, and they experienced stress as a result. Simple routines such as bedtime rituals, snuggling, reading a book, story time—whatever your child finds soothing and can count on—help to build a strong brain-body platform. Mealtimes provide another opportunity to build in predictability without additional time involved. We can add stress-busting goodness when we combine predictability with relaxed, joyful connection and conversation around a dinner table. Of course, life can't be predictable all the time—and we wouldn't want it to be, since we build resilience by responding to challenges and changes. It's possible to have predictability and flexibility at the same time. When life presents a challenge to the predictable and you show your child that you can manage the sudden shift, the child also gets a lesson in resilience.

Randy's Family

It was losing the predictable things in his life that had caused challenges for Randy, who had struggled so much after relocating cross-country with his parents, Lester and Heather, and his sister. Formerly an excellent sleeper, now he required reassurance in the wee hours nearly every night. Previously well adjusted, now he was habitually vacuuming and straightening his sister's toys. These kinds of *control-seeking behaviors* often indicate that a child's basic sense of felt safety has been challenged.

The move away from everything familiar took a heavy toll on his body budget. His behaviors were indicative of a body and brain valiantly trying to manage his dwindling resources. I knew that the first thing Randy needed was for his parents to help fortify his vulnerable platform. First, for proof of the body's wisdom and the rationale for the treatment plan, I asked his parents to closely observe Randy for signs of stress in his body when he woke up at night. I suggested that one of them place a hand on Randy's back or chest to see if they could detect a racing heartbeat or to gently hold his hand to see if his palms were sweaty. Sure enough, they observed both. His body was stressed out, working overtime to find balance again.

I explained to Lester and Heather that Randy's behaviors could be viewed as adaptive: Humans thrive on routines, so when the move forced Randy to lose his predictable and familiar surroundings and activities, his platform was compelling him to seek out safety. Now that he was away from his old friends, home, school, and community, his behaviors were evidence of his system seeking to counteract his anxiety. *I explained that since we are social creatures, we can help our children feel safe; we do this through our loving and thoughtful relationships.*

Randy's need to seek out close contact with his parents, even in the middle of the night, was a tribute to a nervous system doing its job, meeting his need for safety. When children sense threat, it's actually healthy and desirable for them to move toward their loved ones, their attachment figures. *When we feel unsafe, the most adaptive thing humans can do is seek refuge in relationships and reassurance with those we trust in order to feel safe again.* Of course, sleepless nights aren't in anyone's best interest, so we needed to find a better solution, but they were understandable given how the move had upended his life in so many ways.

As I told his parents, there was good reason for Randy's vacuuming as well. *When children feel out of control, they often seek to control simple parts of their life that they can control.* Seeing the little flecks of dirt on the floor disappear into the vacuum cleaner felt satisfyingly predictable to Randy. Doing so sent temporary signals of relief, letting the child focus on *something* rather than just feeling distressed. Randy's vacuuming was also answering his body's need for cues of safety—another signal that Randy needed additional support.

Understanding their son's quest for safety, Lester and Heather came to have more empathy and compassion for Randy, and less worry. They came to view his behaviors as part of their son's subconscious quest to feel safe. Using the two-step method (address the threat and add cues of safety), together we devised a plan to help Randy calm his safety-detection system.

ADDRESS THE THREAT: Heather and Lester patiently empathized with Randy about all the changes, asking him about what he missed most about their old home. His answer: his friends, his teacher, and the room he shared with his sister. He sheepishly admitted that he didn't really like having his own room; being alone at night made him feel scared.

ADD CUES OF SAFETY: Sensing his parents' openhearted desire to understand him, Randy opened up and had plenty to share. At one session, I asked the child what might make him feel more "at home" in the new house. Ultimately, he asked if he could share a room with his sister again. When his parents (and sister) happily agreed, his face lit up and he suggested with excitement that the other room could become their playroom.

I also suggested that the family spend increased calming time together during their bedtime routine, reading books or doing other quiet activities in the hour or so before bedtime. Within a week of the changes, Randy was back to sleeping through the night again. A huge victory for all! He also joined a soccer league and quickly began making friends. After a couple of months, Randy asked if he could have his own bedroom back. The increased cues of safety had proven effective to help him relax and grow into a greater sense of independence.

As Randy's parents learned, another significant benefit of understanding the fundamental human need for safety is that it helps us to be less judgmental and less fearful of our children's behaviors. Instead of blaming the child's "will," we come to appreciate how meaningful children's behaviors are on a brain-body level. We encourage self-reflection and an appreciation of emotions. *We begin to see children's behaviors as protective reactions rather than judging or pathologizing them.* In doing so, we help give our children a strong sense of self, grounded in a respect for their body's reactions, rather than being harsh self-critics.

When we view our children's behaviors through the lens of relational safety and understand how the body's threat-detection system works, we see our children with fresh eyes. We can learn a great deal when our children's (or our own) behaviors shift from *receptive* to *defensive*, or from being calm and agreeable to disagreeable, ramped-up, or out of control. This information helps us direct our efforts to investigating beyond a child's behaviors to what sets them off in the first place. It opens new ways of viewing our children's behaviors through a lens of safety seeking with compassion and empathy. *This deeply felt sense of safety is a substrate of mental health for all human beings.*

THE BOTTOM LINE: The perception of stress is subjective and unique to each of us and determines if a platform is sturdy or vulnerable. Neuroception, or what I call the safety-detection system, is the way the brain and body work together to keep us safe and ensure our survival. With this understanding, the next question is *where* the sense of safety or non-safety is landing for a particular child. In the next chapter, we'll explore how you can learn to deconstruct what your child's ever-shifting behaviors mean in their nervous system.

As we will see, there are pathways in the nervous system that influence children's behaviors depending on the amount of safety and hope they feel (or don't feel) in their brain and body. Now that we understand how safety figures into the balancing of the body budget, we'll discover more about how to gather information by reading your child's behaviors and nonverbal cues to guide your parenting decisions.

RESILIENCE-BUILDING TIP: Appreciate children's behaviors and emotions as a reflection of their subjective detection of safety, challenge, and threat. Humans need to feel loved and safe. There is no greater gift we can give our children than meeting these two essential needs, which will help form the basic foundation of resilience for years to come.

3

The Three Pathways and the Check-In

HOW UNDERSTANDING THE BRAIN AND BODY CAN HELP US RESPOND TO OUR CHILDREN

Our body is always doing what it thinks is best for us.

—Dr. Stephen Porges

Like most parents, I often felt perplexed when one of my children was defiant, not listening to repeated requests or shoving a sibling. What had triggered the behavior? Should I use discipline? Discuss the consequences of the behavior? Ignore the behavior and hope they'd grow out of it?

As a psychologist, I was familiar with many schools of thought, many different approaches, but often that just left me more confused.

What finally helped me to clarify my parenting decisions was coming to understand how the body and brain interact with each other to create the responses we see manifested in our children's behavior. Instead of seeing my children's challenging moments as affronts, I learned to appreciate the occasional outburst or meltdown for the information they provided about my child. When I came to understand how adaptive behaviors are, I grew more confident in my parenting instincts.

Some of that confidence came from understanding the concept we discussed in Chapter 2, neuroception—the way our nervous system detects and interprets threat. But there's more to it: What our brains and bodies *do* with that information can vary based on how it lands in our nervous system. In this chapter, we'll examine three main "pathways" of the autonomic nervous system and how understanding them can help us be even more intentional and directed in our parenting. Let's start with a typical behavior challenge: the struggle over transitions. Children often have difficulty when we ask them to switch from a self-directed activity to something we want them to do. But there are ways to help ease those transitions.

WHEN CHILDREN PUSH BACK

Lucas was eleven years old when his parents noticed a change in his behavior. A gifted student who rarely misbehaved, he suddenly began resisting coming to the dinner table. Most days, his father would pick him up from his after-school activities. Lucas would do homework, then spend half an hour playing video games on the family computer, wrapping up—sometimes reluctantly—when it was time for dinner.

Then one day that reluctance became animosity. When his parents called him to the table, he refused to come, yelling, cursing, and storming out of the room. Unsure what had triggered the abrupt change, his parents tried asking what was wrong, to no avail. In an effort to encourage him to comply, they created a behavior chart, promising rewards if he could manage his time better. That didn't help, either. They warned Lucas that if he didn't shape up, they would take away his privilege of playing the game, but his extreme reactions continued, unabated. Finally, the family contacted me for help.

As challenging as Lucas's resistance was, it was just the kind of behavior that can offer parents useful insights to guide their parenting decisions. In this chapter, we'll examine how to make sense of what we see in our children's bodies and behaviors and in our own nervous systems to yield valuable clues. We'll see how remarkably adaptive humans are and how we can optimize our child's platform to help them face challenges, large and small.

BEHAVIOR AS A GUIDE

We can learn so much from simply observing a child's behaviors. As we have seen, the nervous system is constantly making sense of a great deal of information. Remember that the child's brain and body are constantly reading the internal and external environment and interactions with other people, and then launching behaviors according to that reading. In simpler words, we are always hearing, seeing, moving, smelling, tasting, touching, and taking in sensations from deep *inside* our bodies. In the previous chapter, we learned about *interoception*, our body's internal sensations. In all of these ways, our bodies make sense of the wonderful everyday pandemonium outside and inside our bodies that results in how we feel and behave.

The general sense of how you feel ("affect") has two main features: the *feeling of pleasant or unpleasant* (known as "valence") and the degree of *calmness or agitation* (known as "arousal"). How a person feels is "always some combination of valence and arousal." We gain valuable information by observing how agitated or calm a child is and if that agitation or calmness is experienced on a continuum of pleasantness or unpleasantness. A crying child who is throwing their dinner off the table is experiencing high arousal and unpleasant valence—in other words, high levels of distress. At another time, that same child might be dancing around the living room, energetic but not distressed, as they experience a positive, pleasant valence and high arousal.

Behaviors also provide a rich indicator of the child's body budget balance, and one of the most important things we do for our children in their early years is to help them balance their body budget through our loving interactions. We want to become keen observers in order to discover why the child is exhibiting a behavior—what need that behavior represents in the child's body. So how do we determine the cost of experiences on our children? We observe them to make an educated guess about the state of their nervous system and to discover how much energy they are expending to stay calm in their bodies.

SUPER SIMPLE NERVOUS SYSTEM 101: PHYSIOLOGY

The human body has several "nervous systems." Most of us are familiar with the *central nervous system*, which consists of the brain and spinal cord. We also have the *peripheral nervous system*, which contains the *somatic nervous system*—involved in the movement of our skeletal muscles—and the *autonomic nervous system*. The autonomic nervous system's job is to automatically regulate the internal organs, such as blood vessels and sweat glands, and their functions, so that our bodies are able to maintain homeostasis. The autonomic nervous system, as the name implies, is automatic, and not under our voluntary control, and responds to our perceptions of safety and threat, directing us to degrees of action depending on that reading. Finally, the autonomic nervous system is divided into two major branches: *the sympathetic and the parasympathetic*, which have different effects on our organs. This chapter focuses on the autonomic nervous system, which provides you with information for the road map to customize your parenting to your child's (and your own) brain-body experiences.

We are accustomed to viewing behaviors as either "good" or "bad," children as "well behaved" or "poorly behaved," "polite" or "impolite." Of course, children are much more complicated than any of these dualities. Not all behaviors are willful or voluntary. If you have ever suddenly exploded at your child or your spouse, you understand that. But have you ever wondered *why* you or your child lost control? As we have discussed, that happens when our brain and body detect high levels of challenge or threat, causing people to move and act in ways that they typically wouldn't when feeling safe and in control. It's critical to appreciate the difference between *purposeful misbehavior* and behaviors driven by a sudden shift in the autonomic nervous system. This understanding helps us ground our response according to our child's platform.

The Brain-Body Pathways That Influence Behaviors

According to the Polyvagal theory, our body responds to our moment-to-moment experiences in order to keep us safe, through three pathways of the two branches of the autonomic nervous system, the *sympathetic* and the *parasympathetic* branches. Two pathways are found in the parasympathetic branch, the *dorsal vagal pathway* and the *ventral vagal pathway*. The third pathway is called the *sympathetic nervous system*. Each pathway instinctively—that is, automatically—directs our body's internal responses and behaviors according to the degree of threat or safety we detect at any moment. Each has its own range of receptivity and accessibility, its own spectrum from openness to defensiveness. Understanding these pathways, and on which of them you and your child are located at any moment, is essential to responding appropriately and supportively to your child's needs. And don't worry about memorizing these scientific terms! In the coming pages, I'll offer an easy shorthand to help you remember the key concepts.

Since all humans—adults and children—have these pathways, they are useful in examining our own reactions as well as our children's. Our most important tool as parents is the power of observation. Often, we act *before* considering the underlying meaning of our child's behaviors, focusing on managing or correcting the behavior instead of reflecting on what the behavior means and what clues it holds. *When we tune in as nonjudgmental observers, we can gain a new appreciation of our child's behaviors and move away from impulsive parenting decisions, which can often turn out badly.*

We can observe, for example, how quickly and urgently children move their body (or mouth) and make inferences based on their tone of voice, muscle movements, heart and lung activity, body gestures, and behaviors, which provide valuable signals we can read and glean to direct our parenting efforts. We can make educated guesses by looking for behaviors that cluster together, providing clues about the state of a child's physiology (what I've called the platform).

We'll now examine the three main autonomic pathways that protect us and how we can use that information to help guide our parenting decisions. We can use what we see in a person—their rate of movement in the body, their expressions and gestures, and the tone of their voice—to infer the *state of their autonomic nervous system.* Every child is unique, and we need to discover the signs that reveal the state of each individual child's nervous system. In short, you're getting to know your child (and yourself) in a new way, from the inside out.

In the not-too-distant future, we'll have technology to measure our autonomic physiology just as a Fitbit or smart watch measures heart rate. Researchers have developed *wearable sensors* that detect metrics such as variation in the interval between heartbeats (known as HRV, or heart rate variability) and changes in skin conductance due to sweat (EDA, or electrodermal activity) that provide information about activation of the autonomic nervous system. A company called Empatica has already created the first clinically validated and FDA-approved device that provides valuable insights for individuals with epilepsy and their caregivers.

But we certainly don't need sophisticated technology to know how to support our children—we simply need to be present and observant in our interactions with them. So let's look at all we can learn from the color pathways, starting with the green pathway.

The Green Pathway: Safe and Secure, Open and Receptive

Let's start with the ventral vagal pathway of the parasympathetic nervous system, which the Polyvagal theory describes as the *social engagement system*, which we'll simply call the *green pathway*. In this pathway, a person feels safe and social, connected with others and with the world around them. When the body perceives safety, we're in the green pathway.

This pathway supports calm in the body, allowing us to connect readily with others. When we're in the green pathway, we send others signals of connection and communication. This pathway gives rise to a child's learning and growth, and our own best parenting, because it supports our ability to experience joy and play, to think, to plan our actions, and—when we reach a certain developmental ability—to control our emotions and behaviors.

When a person is in the green pathway, our behaviors provide clues as to the state of our nervous system. The green pathway reflects a healthy body budget and is one of receptivity, where we and our children are most open to being together in positive ways. Of course, the words and behaviors I describe here are meant as a general guide and comport with what I've observed in my clinical work. You should be alert not for single words or behaviors but descriptions that cluster together that will help you better understand (and make an educated guess) about the degree of activation of your child's nervous system.

Words that describe humans in the green pathway:

- *Safe and secure, calm, content, happy, joyful, cooperative, playful, attentive, alert, focused, receptive, open, peaceful, engaged*

*In our **bodies**, we might see:*

- *Focused attention*
- *Relaxed posture without clenching or gripping*
- *Regular and rhythmic breathing and heart rate*
- *A variety of tones of voice (not monotone)*
- *Appropriate and balanced body reactions (with movements neither too fast nor too slow)*
- *Smiles, neutral or relaxed facial muscles*
- *Alert, bright, or shiny eyes*
- *Giggles or other joyful expressions*

When our children are in the green pathway, they are receptive to us and to their surroundings. In this pathway, the child feels safe, open, and available. Children enjoy their play and are open to trying and learning new things. *If you want to encourage your child to stretch beyond their comfort zone, this is the best time to try it: when the child's platform is in this place of maximum receptivity.*

When parents are in the green pathway, we are most likely to trust our gut instincts and be present and patient with our child. We are better able to control our emotions. Instead of being driven by worries, manic thoughts, or a racing heart, we are able to make thoughtful decisions. We feel more hopeful, playful, available, positive, and encouraging with our child and others. We're more inclined to socialize and be around other people (if that's something that feeds you). Remember: It's all about appreciating our individual differences.

The Green Pathway Supports Safety, a Gleam, and Joy

Warmly attuned and sensitive parenting builds the child's platform from the ground up, and its benefits continue throughout childhood and beyond. Children feel safe in this pathway, which naturally propels them to do what children love to do: communicate and play.

Think of a newborn gazing into your eyes, or a one-year-old discovering the joy of walking and looking at you for your smiling support. Or a preschooler who draws a picture and proudly shows it to you. Or an eight-year-old who spontaneously tells you about a problem at school while you're on a walk together. These things happen when a child feels safe and the green pathway is supporting social engagement.

Stanley Greenspan, the child psychiatrist, described children and adults engaged in joyful play as having a "gleam in the eye." When you see your child of any age with that gleam, a calm look or smile on their face, and a body that's not moving too fast or slow but poised to play, you know they're in the green pathway. And there's a good chance that when your child has that gleam, so do you. We experience joy, safety, and connection in this pathway.

ACTIVITY: Think of some of the coziest and most pleasurable, joyful moments that you've shared with your child. Focus on the feelings evoked in those memories. What kinds of activities or circumstances generate moments when you or your child have gleams in your eyes or a quieter feeling of connection? (You may find it difficult to recall such moments, and if so, that's okay. In subsequent chapters, we'll discuss how to make them more frequent.)

ALL PATHWAYS ARE ADAPTIVE, BUT THE RED AND THE BLUE ARE MORE COSTLY

Of course, nobody lives permanently on the green pathway. We humans are reactive, instinctive creatures. Life is constantly changing, unpredictable, and full of obstacles and challenges to which we must constantly respond. So it's not useful to think of the green pathway as "good" and others as "bad." *All of the pathways are adaptive.* As we encounter dilemmas and challenges, we can expect to cycle in and out of each pathway throughout the day, even though the energy expenditure associated with the red (highly activated) and blue (immobilized) pathways is more costly to the body budget. What's important is that we determine when our children need our help finding their way back to the calm stability of the green pathway, and when it's better to stand back and let them find it on their own. It's a balancing act, but let's learn a little more before I share the formula for discovering how and when to help.

The goal is to have a *regulated* nervous system, one in which we are able to recognize when we have left the safety and connection of the green pathway and can find ways to come back to it from the other two pathways of the autonomic nervous system, as we will now discover.

The Red Pathway: Move It!

When our safety system detects too much challenge or threat, we instinctively and automatically move from the calm of the green pathway to the more protective red pathway. In brain science terminology, a *biobehavioral* reaction stirs us to protect ourselves from detected threat by taking action. That almost always means a kind of *movement*—from moving your mouth to moving the whole body. Some examples: yelling angry words with pressured speech, hitting, shoving, or even running away. In order to feel safe, humans who detect threat are activated on the inside and feel the need to *move*.

When our nervous system detects threat, we shift from the green pathway, and in the process we can lose control over our behaviors and emotions. Automatically, we enter the red pathway, called the sympathetic nervous system, which prompts *fight-or-flight behaviors*—the sort Lucas demonstrated when he began to curse loudly and stomped out of the room in frustration over his parents' request that he stop playing his video game and come to the dinner table.

Words that describe humans in the red pathway:

- *Angry, aggressive, hostile, disruptive, noncompliant, defiant, misbehaving, having a tantrum, hyperactive, hostile, argumentative, pressured*

*In our **physical bodies**, we might see:*

- *Intense, narrow focus or continual roving with scattered attention*
- *Running away, constant motion, or an increased need to move or escape*

- *Fast, erratic, or impulsive movements*
- *Hitting, attacking, kicking, spitting, jumping, or throwing objects*
- *Shallow, fast, or irregular breathing patterns*
- *Increased heart rate*
- *High-pitched, loud, hostile, gruff, or piercing tone of voice; out-of-control laughter*
- *Eyes tightly closed or wide open*
- *Tense, clenched facial muscles or jaw*
- *A range of facial expressions or a forced smile*

A child in the red pathway is not receptive to reasoning or a wide range of requests and typically displays out-of-control behaviors. In mild cases, a child experiences a moderate challenge and may simply start to fuss, whine, complain, or refuse to do what's asked. In more extreme examples, they might tantrum, hit another child or adult, or run away. Often, the change in mood and behavior happens very quickly, as the pathway can shift from green to red instantaneously.

Most of our difficult parenting moments happen on the red pathway. This is where we are suddenly triggered and "lose it," doing and saying things that we later regret. In the red pathway, something harsh, dismissive, or hurtful might slip out of your mouth before you know it and you may feel compelled to discipline severely. You might do or say something that is out of character or begin to feel physical symptoms like a wave of heat throughout your body, a racing heart, sweaty palms, or a gripping in your stomach. The neuroscientist Bessel van der Kolk summed up the phenomenon in the title of his book *The Body Keeps the Score*. You experience the feeling of stress in your body and mind at the same time.

The red pathway is the pathway of "mobilization" that helps humans escape dangerous situations, such as being preyed on by animals, by either fighting or moving—quickly. (Hence the phrase "fight or flight.") This is the pathway that takes over when the safety-detection system sends a signal that we need to overcome a threat. Whether the safety-detection system is triggered *depends on the individual's unique reaction*, so there are times when a child might experience a fight-or-flight behavior even when they are objectively safe, as was Lucas. That was also the case for Randy, from the previous chapter, who suddenly had difficulty sleeping in his own room after the family move. His mind knew he was safe, but his safety sensor took a while to register that reality. Our marker for knowing how to support a child isn't necessarily our objective assessment of a situation, but *how the child's body is reacting to it, and the toll it's exacting on the child's body budget*.

ACTIVITY: Think of a time when you or your child (or both) landed in the red pathway. Try to recall what your child's body and facial expressions looked like during that experience. Now try to remember how it felt in *your* body as you were trying to manage it. Don't linger here for an extended period, just long enough to recall the physical memories of it, and try not to judge your child or yourself. Observing our body's reactions is essential to helping your child get back to the green pathway, and simply calling to mind these moments is the first step in knowing what to do when these sensations and feelings surface.

We all have moments when we think, do, or say things that are a product of landing in the red pathway. It's a part of being human. We all find ourselves in situations where we are triggered and lose control of our emotions or behaviors. The key is to compassionately recognize what is happening and adjust course to find our way back to the green path as soon as is practical and possible.

Red-Pathway Behaviors: Not "Bad" but Protective

It's important to remember that when a child is squarely in the red pathway, they're not likely to modulate or control their behaviors very effectively. Rather, body and brain are trying to protect the child. The resulting behaviors may appear negative, but they're also protective from this perspective.

Understanding this helps us shift how we perceive children's severely disruptive behaviors. Instead of viewing red-pathway behaviors as "bad," we can see them as signaling *vulnerability* in children, helping children to protect themselves through *instinctive* rather than *willful* or rude behaviors. These are the self-preserving body-up behaviors described in Chapter 1. When a child "goes red," we need to adjust our parenting techniques; in the red pathway, children can't think and function well because they are highly agitated. This is why punishing a child in this pathway is counterproductive. In the red pathway, the child is not receptive but defensive. A child in the red pathway is expending the resources of their body budget at a high rate. This pathway is costly and serves a purpose. *But that purpose isn't, as many educators and those in my field who focus on behavior management assume, to get out of something or get something.* It's staying safe—and surviving. Punishments will only send the child deeper into the red (or possibly into the third main pathway, which we'll describe later in the chapter).

To reiterate, the trigger that sets off a child's safety system isn't always genuine danger in the environment. Sometimes the child's nervous system registers something harmless as threatening in the moment. *As we saw in Lucas's case, sometimes a child registers a reasonable request—like stopping an activity to come to dinner—as a threat.* Even though the child is objectively safe, their body goes into an activated state in which we can't generally reach them through talking and our logic.

There are body-based reasons for this reaction. Physiologically, it's difficult for children in the red pathway to distinguish the sounds of the human voice. When a child is in the red, with the sympathetic nervous system in full force, the middle ear muscles shift away from distinguishing the nuances of human voices, to hearing low-frequency and predatory sounds. This explains why, when children—or adults—are in a highly activated state, they often seem not to listen. Their ability to hear the human voice is compromised. In this triggered state, humans can also misread facial cues. When a child is in the red pathway, they might register a neutral facial expression as an angry one, their safety-detection system activating their defenses. That's why when Lucas "went red," his parents couldn't reason with him or discuss the situation. His body was poised to move, not to reason or listen.

Look again at the list of words that describe the red pathway behaviors. Many can be easily construed as intentional "misbehaviors." When we assume a child has acted within their conscious control, most parents are inclined to discipline or otherwise instruct the child. Our instinct is to correct the child as quickly as possible. We don't want our children to misbehave, and we want to raise them well.

I certainly felt that way one of the first times I experienced my own child's red-zone behaviors. I was relaxing at a relative's festive birthday picnic when I saw my three-year-old suddenly bite her five-year-old cousin on the shoulder. Aghast and embarrassed, I sprang to my feet and shouted at my daughter, causing her to burst into tears. I felt perplexed, certain that she knew that she shouldn't bite another child and not understanding why she had.

What I didn't realize was that she—and then I—had landed in the red pathway. The biting wasn't intentional but rather a default result of her safety system detecting threat. At the time, I didn't understand that my child's nervous system subconsciously "chose" this behavior, which wasn't a preplanned misbehavior but rather an automatic *stress reaction*. What I didn't yet realize was that her body had a tendency to strongly overreact to shifts in her environment, and to certain sounds and volume. So the commotion at the first big birthday party she had attended simply overwhelmed her green pathway, landing her in the red. The result? Her body attacked the nearest thing, which happened to be her innocent cousin. Seeing her do that sent me onto the red pathway, too. Had I been aware of the brain-body connection back then, I wouldn't have yelled at her, shaming her and also causing her safety-detection system to detect an even higher degree of threat. We were both a mess.

We enter a new realm of helpful parenting strategies when we come to understand that many red-pathway behaviors reflect a child's vulnerability and a protective fight-or-flight response, not willful disobedience, and signal the child's need for help, not increased discipline.

The Blue Pathway: Disconnecting and Withdrawing

While the red pathway is associated with movement, the *dorsal vagal*, or blue pathway, is the opposite. Overwhelmed, a person conserves energy by withdrawing from connection and contact with the world. When a person is in the blue pathway, we can see, hear, and feel this lack of contact and engagement. Occasionally, Lucas ended up in the blue pathway. He told his parents he felt "spacey" and would lie on his bed for hours, not wanting to connect or answer a question.

Some words that describe people in the blue pathway:

- *Sad, slow, blank, distant, disengaged, flat, frozen, absent, disinterested, disappearing, hopeless*

In our **physical bodies**, we might see:

- *Slow or few movements, slouching, wandering aimlessly*
- *Apparent drowsiness or the appearance of being checked out*
- *Little or no exploration, play, or curiosity*
- *Slowed heart rate and breathing*
- *Moving slowly or even appearing to be immobilized*
- *Speaking in a flat voice with little intonation and/or cold, soft, or sad sounds*
- *Eyes that are glazed, turned downward, or not seeking contact with others*
- *An expressionless face, with no smile*

We all experience moments of disconnection or checking out. Sometimes, for good reason, a child wants to be left alone, find quiet, and recharge. We can expect these behaviors because, as we've seen, they are all adaptive to the child's shifting platform, and we all have different needs for solitude. Since all of the pathways are adaptive, most of us will occasionally be on the blue pathway, but we won't stay there. At the extreme, however, the blue pathway signifies that a person's nervous system is detecting very high levels of threat and is protectively conserving energy. In this pathway, children and adults can feel empty, depressed, hopeless, or lost, and they need big deposits into their body budgets.

It's important to note that people don't always associate this pathway with stress because the child might appear overly self-sufficient or calm—they may seem green when they're actually blue. The most basic way to tell the difference between a calm child and a child who is in the blue pathway is whether or not they are engaging with you, exploring their world, and playing.

Most people occasionally feel down or momentarily stuck, but we should become concerned if a child stays disconnected for long periods of time and seems stuck there. This is a sign that the child needs additional support to experience solace and anchor in human connection again.

When parents are in the blue pathway, we might feel empty, disconnected from others, foggy, checked out, unable to think or act, or even immobilized. It's a sign that you need to do something to feel connected to yourself and others soon, and that your body budget is significantly overdrawn. Our children need us to have a mind that can think and a body that can respond to their needs. If you notice that you or your child is disconnecting or feeling hopeless or lost for weeks or months at a time, it's important to seek professional support to help you discover ways to reconnect with the most important power source humans have: each other.

ACTIVITY: Think of a time when you or your child felt disconnected or "blue." Do you recall the feelings you had in your body or thoughts in your mind? It can be difficult emotionally to consciously recall this kind of suffering, but doing so even for a moment will help you recognize what to look for. Feeling this way is part of the human experience, but fortunately for most people it's not where we live.

Lucas, for example, rarely withdrew for long. His platform generally chose actions—like yelling and running to his room—rather than stillness. Red behaviors are difficult to manage as parents, but even if frustrated, the child remains engaged with others; blue-pathway behavior looks more like giving up. The next time you see your child "losing it" in the red pathway, it might be helpful to admire the way the child's nervous system responds to stress so robustly. *Instead of seeing disruptive behaviors as "bad," we ought to appreciate what they tell us about a child's nervous system actively responding to perceived stress and expending a great deal of energy as they try to cope.*

Mixed or Blended Pathways of Experiences

While the distinct color pathways are helpful descriptors of the three states of the autonomic nervous system, the reality is, of course, more complex. Some researchers are looking into how the various pathways blend or overlap. A meditative state, for example, represents a blend of the blue and green pathways: A person might be relatively immobilized, but also feeling safe, so meditation is an example of stillness without fear. We know from research that this still, safe state is healthy and reduces stress in the physical body: Various forms of mental stillness in the pathway of safety include mindfulness, prayer, and yoga.

Play, too, likely involves a blended state, in which the green and the red pathways work together. In Chapter 8, I'll describe how play can provide a brain exercise and a way to help children resolve emotional challenges. *Additionally, a child who looks quiet or frozen on the outside might be activated on the inside, with an increased heart rate and other characteristics of the red pathway working internally.* I have observed this state in many children who display varying levels of anxiety and hypervigilance in school. They may appear to be "good" students, but inside they are quite activated and unsettled, and as such have vulnerable platforms even if they appear to be fine. Such children might only display disruptive behaviors at home, for example, and parents are puzzled by their child's "excellent" behavior reported by the teacher at school.

Sometimes, even though parents know that their child isn't feeling safe in their body at school, teachers are surprised to learn this.

These hypervigilant "mixed"-pathway children fly under the radar because they can look so compliant. We shouldn't mistake "well-behaved" or compliant children with green-pathway, safe-feeling children. The child might be detecting threat but is unable to talk about it or show it. They are challenged on the inside, but we don't see it. We need to look closely at the whole child, at their face, tone of voice, posture, and desire to play. It's also useful to talk with children and help them to feel safe enough to let us know about the sensations they feel in their body.

While it's not necessary to understand the complex nuances of blended autonomic pathways, simply knowing that there are likely combinations can help us better understand what our children need. Researchers will discover more in the future, but for now, what's important is to get to know your child's stress triggers. When we see strong indicators in the red and blue or combinations thereof, that's a sign that the child needs our supportive presence to help find a better challenge zone. The child may need more infusion of your emotional strength and stability (or an adjustment in their circumstances) to help them build their flexibility and capacity to manage stress.

BE PREPARED FOR ANY AND ALL PATHWAYS

Reframing "negative" behaviors as outward reflections of a child's physiology allows us to have more compassion. We can shift our focus from the frustration of dealing with challenging situations to appreciating what our child requires—and indeed, what we require—to feel safe. When we understand that the brain's main job is to maintain the body budget, and that this survival-based process underlies *all* of the pathways, we can view them all with equanimity. However, it's important to know that the red and blue pathways are more costly to the body budget, and we don't want children spending too much time there.

We recognize that human nervous systems, our children's and our own, are fluid and dynamic, and are expected to cycle through various pathways, even though it's more likely that we all get our best life's "work" done in the green. Understanding that these pathways all serve a purpose in our nervous systems can help us to understand why we sometimes do or say things we regret and feel guilty about as parents. Sometimes, our instincts lead us in directions that don't feel right. *If you feel you're being reactive rather than intentional, have compassion for yourself. Then look beyond your own behaviors to what might be triggering your explosive reactions.* I'll walk us through understanding our triggers in Chapter 5. Parental guilt can be devastating, and feeling bad about ourselves certainly doesn't help us to be better parents or build up our body budgets. My hope is that gaining an understanding of the brain-body connection will alleviate guilt and empower you to be better prepared for challenging parenting moments.

PRACTICE: When you wake up in the morning, take a read on the color of your pathway. Are you feeling anxious and stressed (red), disconnected (blue), or pretty good and ready for the day (green)? Don't judge your state but observe it. Recognizing the state of your nervous system without judgment is the first step. It can help you plan your next hour: What does your body need? What are practical ways you can meet those needs?

Using the Colors to Measure the Child's Body Budget Balance

We can use the colors to help us determine the stress load our children are carrying and help them discover their optimal challenge zone. We help children who are experiencing too much stress—whose systems are in the red and blue too often or for too long—by making deposits from our own nervous system into their nervous systems, through our loving interactions. If a child is running low in their body budget, then it's a signal for the adult to provide support, not discipline or even teaching. And it's *not* a good time to ask them to take on a new challenge or to learn something new—either of which would be metabolically costly.

Ideally, we want children to spend the majority of their waking hours in the green pathway, rather than in the defensive, fighting, or escaping mode (in the red) or in a place of disconnection (in the blue). If a child is spending too much time in the red or blue or mixed shades, we can intervene with connection and support, to help get back to the safety of the green pathway as soon as is practical. It's helpful to tabulate how much time each child spends in the pathways.

As a reminder, the shorthand for the pathway colors we just discussed are:

- **GREEN:** *Calm, alert, cooperative*
- **RED:** *Moving, fighting, or running away from you*
- **BLUE:** *Disconnecting, losing contact, not communicating, possibly shutting down*
- **MIXED RED AND BLUE:** *Hypervigilant, anxious, might look calm on the outside but is activated and unsettled on the inside*
- **ANY AND ALL OTHER COMBINATIONS** *you may infer from observation*

In addition to determining what color pathway your child is on, we can measure three factors that provide us with more detail about how their brain and body are managing life's demands. We can tabulate *how often, how intense, and how prolonged* your child is in a state of stress.

Why? Intensity, frequency, and duration are all important considerations in our assessment of a child's level of distress and provide clues about how we can help the child. Consider the difference between a fussy, whiny toddler and one in a full-blown tantrum. What varies is how much subjective distress the child's nervous system is experiencing. A whiny toddler is likely not in a full-blown red pathway but maybe a pink or light red one. A fussy toddler's body is taking a hit and perhaps beginning to fatigue but hanging in there. The child is not in a state of deep distress. But once that toddler, or anyone for that matter, begins to lose control and is red-faced, screaming, falling over, and inconsolable, that's a higher level of distress and more costly to the child's body budget.

ACTIVITY: We can look at the patterns for infants, tots, and older children to track *what activities or circumstances prompted the different pathways, when it was happening, how long it lasted, and how intense it was.* This will help you understand how to tailor and titrate your support for your child and better understand what's going on under the tip of the iceberg. Make a weekly journal that has three or four columns for the green, red, and blue pathways, and any combinations you observe.

Date/Day of the week:

Time Began:

Time Ended:

What was happening?

What color pathway was the child?__ What was the level of distress from 1 to 5? (1 is mild distress and 5 is extreme distress.)__

What color pathway were you?___ What was the level of distress from 1 to 5?__

How much time a child spends in each of the pathways will vary according to the child's age and stage. Of course, infants will have more time in the red and also more time asleep, as they are completely dependent on us to help regulate their body budgets, and toddlers are all over the map developmentally. While physiological research needs to be conducted to establish further guidelines, for children ages five and up, we expect that during the waking hours, about 30 percent of the time at a maximum should be in the red, blue, or blends of the pathways according to the context.

MISBEHAVIOR VS. VULNERABLE PLATFORM BEHAVIORS

When we use the pathways of the nervous system to see children's behaviors in a new way, we start noticing the difference between willful misbehavior and behaviors that signal a child's vulnerability and need for support. Our culture generally interprets children's behaviors as either "good" or "bad," "compliant" or "noncompliant." Using this limited lens leaves us few options besides trying to control the child or punish the behaviors.

The problem is that punishing or controlling the child doesn't make the child feel safe. *Distinguishing between a stress behavior arising from deep within the nervous system and purposeful misbehavior can help us be more compassionate to our children.* Understanding the difference opens a new array of parenting options to help with challenging behaviors.

Of course, when a child tests limits, does something dangerous (as when a toddler reaches for an electrical outlet), misbehaves intentionally (sneaking a phone to school against the rules), or needs correction, we ought to provide clear and loving boundaries. We are our children's primary teachers and guides. But when a child's behaviors stem from the red or blue pathways, the platform is vulnerable, and our first priority should be to help the child back to the green pathway by connecting and not punishing. The hope is that viewing a child's behaviors through this brain-body lens will help you know when to try to stabilize your child and when to hold the line, when your child needs teaching and redirection and when they require a reconnection to their sense of safety and trust with you.

THE CHECK-IN IS THE NEW TIME-OUT

As my now-grown children will attest, I used time-outs as much as any parent. Decades ago they were the pride and joy of behavioral methods. But after I studied the nervous system and relational neuroscience, I realized that this technique doesn't optimize a child's or parent's platform, nor does it help a child move from the red or blue pathway to the green.

The time-out reflects a parenting approach that targets surface behaviors rather than underlying causes. It's based on the assumption that a child who has just misbehaved (and may be deeply upset or agitated) is receptive to teaching and will somehow learn from being isolated. Of course, we now know that the state of a child's platform may make it impossible for the child to learn or benefit from such an experience.

So let's examine an alternative to the time-out: the check-in.

Summary of the Check-In: A Barometer of the Body Budget

1. ***Check yourself*** *by observing your pathway color.*
2. ***Check your child*** *by observing your child's pathway color.*
3. ***Proceed*** *with strategies personalized to your child, by*

compassionately working from the body up to the top down. The goal of this third step is to provide emotional attunement, *tuning in to what your child's nervous system needs in order to come back into balance of the green pathway.*

Parents—or any adult caring for a child—are the key ingredient in helping a struggling child to feel calm, so the check-in starts with checking in with ourselves. We start with this because when we communicate with a child, the child senses safety or threat first on a *nonverbal* level. *How we speak* is initially more important than *what we say*. Most parents know that an out-of-control adult trying to interact with an out-of-control child can lead to disaster, so we start by assessing our own color, which reflects our own level of receptivity or vulnerability. The check-in is useful because when we provide attunement first, we help build the child's frustration tolerance and challenge zone. With time-outs, the likely lesson a child learns is that their behaviors or emotions are intolerable and must stop in order for us to be together. *With the check-in, we go to the heart of the problem: a dysregulated child.*

Step One: Check Yourself (How Do I Feel?)

First, check your own pathway color—without judgment or shame, and with the knowledge that it could be any of the three. Your pathway color is an indicator of your body budget and what you have (or don't have) to give to your child in the moment. Do you feel calm, agitated, or somewhere in between? Try not to judge where you are, just acknowledge what your body already knows.

In the green pathway, we feel safe and secure, and we exude safety and security. That's why we start here. *We share our own sense of calm and safety through the look on our face, the tone of our voice, and the actions of our body.* When you are in the green, you are best able to choose the appropriate reaction to your child rather than exploding or disconnecting.

That doesn't mean you have to be in the green pathway. The goal isn't perfection but awareness.

Take stock: Are you experiencing the green, red, or blue pathway? *Green feels like you are in control and ready to parent thoughtfully.* If that's how you are feeling, then proceed to **Step Two.**

If you're not, you may be triggered, revved up (red), with stress in your body that may include a racing heart or thoughts, sweaty hands, breathing that is fast or shallow, and the likelihood to say or do something negative to your child. Your body is in a stress reaction. Simply name it to yourself. Do you feel like you are in the blue pathway? The signs might include feeling disconnected, sinking, stuck, or frozen. That's okay, too; just name it. Perhaps you are feeling combinations of the various pathways. The key is to determine if you are in control enough to parent. If you're not, that's okay; it's time to stop and observe.

Then pause a moment, take a breath, make sure your child is safe, and consider this question: "What do I need in this moment?" Within the constraints of the situation, figure out what you need to do to regain control, find your way back to the green, and interact positively with your child. Chapter 5 will cover many tools and techniques to help us recenter, regain control, and prevent—or recover from—moments we regret. But for now, I'll share that one of the easiest and quickest methods for many people is taking a deep breath, extending the exhale a bit if you can. (Take a few if it helps, but be aware that this doesn't work for everyone; we are all different.) The general idea is to take care of yourself in the moment. If it's possible, do you need to step away? Perhaps you let the child know you are going to take a moment and come back. Make sure your child is safe, and then find momentary space and perspective. Maybe you take a sip of water, brush your teeth, or walk into another room for a minute or two to recover enough to engage productively.

ACTIVITY: When you have some time by yourself, think about and write down what helps you feel calmer in the moment. Is it a certain kind of breathing? Naming the feeling to yourself? Some type of movement, a hand over your heart or squeezing your toes? Repeating a soothing phrase or mantra?

Step Two: Check Your Child (How Is Your Child Feeling?)

Next, check your child's pathway color. If your child is on the green pathway, then you're clear to try to interact or talk to the child and figure things out together. This works well if your child is older and able to use words to communicate, as the green pathway is where we can talk to the child and begin to solve problems together. (In later chapters, we will explore what to do depending on your child's age and developmental abilities.)

Of course, if you've encountered a moment in which you would typically use a time-out, then the child probably *isn't* in the green pathway. *Are they in the red pathway* (yelling, agitated, hitting, or needing to move)? *Or are they in the blue pathway* (disconnected, disengaged, or not responding to your attempts to communicate)? *Or perhaps a hybrid of the two* (whiny, pleading, vigilant, appearing anxious)? Any of these indicate that your child is vulnerable and needs platform building rather than teaching or consequences. Proceed to the next step then, for ways to help the child back to the green pathway.

Step Three: Proceed with Compassionate Strategies Personalized to Your Child

The third step of the check-in is to choose your response based on the child's pathway, in order to attune with your child and get back to the green together. If the child is in the red or blue pathway, or combinations thereof, the first thing we should do is witness our child's distress without judgment. *This is a profound first step because human beings feel better when their struggles are witnessed with acceptance and love and they feel that they aren't alone.* The simple act of compassionate presence begins to calm the nervous system. Then, begin to figure out what your child's nervous system is telling you. Doing this entails being present for the child—physically, mentally, and emotionally—in a process called co-regulation, which basically means we make deposits into the child's body budget according to the child's needs at the moment. I will describe that process in detail in the next chapter. Basically, this third step gives us a general sense of what to do next.

Sometimes, that next step will involve reasoning with the child if they're ready; sometimes it might be soothing the child in order to reason; and sometimes it might be holding a firm line because you believe the child can handle the challenge. It all depends on the situation, your child's pathway in the moment, and other elements that I'll describe in Part II, Solutions, where I'll apply to infancy, toddlerhood, and childhood what we've learned in the first three chapters. You'll determine a formula based on what works best for your child.

This is the heart of brain-body parenting: to attend to the child's *platform*, rather than to simply address the behaviors themselves. We do this by offering cues of safety, specific to the child's nervous system and delivered through your connection with your child.

Parenting in this way, from the body up, is not only kinder and more tender than top-down disciplinary approaches, but it helps to build and support your child's platform. Once we do this, concerning behaviors often resolve naturally, because we have helped the child out of their defensive (protective) state.

LESSONS FROM LUCAS

With our new understanding of the three pathways, let's return to Lucas, who struggled to contain his emotions when his parents asked him to leave his video game and come to dinner.

As I explained to his parents, we needed to figure out what made Lucas shift so quickly from calm and regulated (green) to fight-or-flight (red). It appeared that what triggered him specifically was making the *transition* from an activity he clearly enjoyed to family time. Children often struggle with such transitions because they have difficulty shifting gears. Some children find it stressful to shift to something new that is not self-directed. What we see as a result is a child with oppositional behaviors, whose body is managing the subsequent feelings and sensations around the cost of that shift. Sometimes a child wants to exert control over a situation and lacks the flexibility to cede control. That capacity fluctuates throughout childhood, depending on each child's emotional development and stress level. It can take years—in some cases all the way to adulthood—for a child to learn how to use their self-control to manage what their body perceives as stressful. *Many adults who struggle with their emotions are on this journey as well.*

After I introduced Lucas's parents to the idea that his body was showing signs of stress through his behaviors, they both confided that they found it difficult to stay calm when they anticipated the strong negative reactions coming from their son. Acknowledging this helped them understand that their own green pathways were going to be a part of the solution. The next step, checking the child's pathway color, was easy, since Lucas went instantaneously to red. That was a clue that Lucas's body budget was running on a deficit and that he was likely managing other stressors beyond his parents' awareness. This was a signal that we needed more information.

His parents told me that Lucas's outbursts had begun a few months earlier, just as he had started a new school year. Sensing that might be significant, I suggested that his dad try picking up Lucas early from his after-school program for a week to spend some relaxed time when the two could talk.

Taking time to play catch at the park and walk their dog with his son, the father discovered that Lucas was having a hard time. He admitted that an older boy at his after-school program had been bullying him. Thinking back, the father realized the bullying had started right around the time Lucas had become so resistant at home. It was no wonder his stress levels were so high, and no wonder transitions had become so challenging for him. Lucas's body budget was being depleted at school, and he was coming home with few internal resources. Armed with this new information, Lucas and his parents met with the school's director, who devised a plan to ease the situation, provide support and safety for Lucas, and address the bullying.

With his stress level and cue of danger properly addressed, we turned to helping Lucas build his ability to transition from the video game to dinner and conversation with his parents. The third step of the check-in is to practice attunement, to recognize what the child's nervous system needs to feel safer, and to build a more solid platform. *Nothing does that as effectively as a caring adult who can stay calm through a child's emotional storms.*

Lucas's parents discussed the problem with him, enlisting his help to find a solution. But Lucas wasn't yet able to discuss his emotions and behaviors. His dad devised a solution: He started sitting close to Lucas while Lucas played the video game, doing some of his own work and occasionally offering encouragement. He also began easing the transition by giving Lucas a ten-minute signal before dinner. In the process, he shared his green pathway and multiple small messages of safety with Lucas. In short, I showed the father how to "lean in relationally" in order to help his son *modulate the intensity of his reactions* to the difficulty of transitioning away from his video game.

The approach proved successful. Within a week, Lucas reduced his protests when asked to turn off his game. To be sure, he still resisted on occasion, often bargaining for more time, but he managed to stay on the green pathway. Now that the school had dealt with the bully, it had reduced the threats that had caused his stress at school. His dad then provided the emotional deposits into his body budget, building up his account. A month later, his dad no longer needed to walk him through the transition; Lucas managed it on his own.

One detail worth noting: The initial plan did *not* include teaching Lucas about his nervous system. That information was to help his parents better understand their child's behaviors. In supporting our children, it's best to avoid using top-down strategies too soon. (In future chapters, I'll show you how to teach children about their nervous system when they're ready and it's more meaningful and useful.)

For now, it's essential to remember that *addressing a child's behavioral challenge (or any other challenge) starts with supporting a child's sense of safety and trust through our relationship. The more physiological distress children are experiencing, the more they need our help.* Don't get thrown off by your child's red pathway behaviors as a call for increased discipline. The engine for a child's development throughout life is a process of helping a child to establish a strong green pathway called *co-regulation*. It's what we do when children are in the subjective feeling of distress. And it's very powerful. We'll focus on what it means and how to do it in the next chapter.

RESILIENCE-BUILDING TIP: Pay attention to signs in your child that provide information about the state of your child's platform. These include the color pathways that represent feeling *safe and calm* (green), feeling *agitated* (red), feeling *disconnected* (blue), or feeling a combination of the colors. The non-green pathways indicate that your child is vulnerable and needs additional emotional and relational support. The pathways of your child's nervous system can guide your parenting decisions, including how you set limits and follow through on expectations—both of which are compatible with empathy and understanding.

PART II

SOLUTIONS

4

Nurturing Children's Ability to Self-Regulate

To connect and to co-regulate with others is our biological imperative.

—Dr. Stephen Porges

In the first part of this book, we learned that the body and brain interact in complex ways that influence our children's emotions and behaviors (as well as our own). Children's challenging behaviors are often a sign that their nervous system is responding to stress. Gaining an awareness of the body budget is a useful tool for guiding our parenting decisions, helping us gauge what our children need in order to feel calmer and more alert.

We also discovered that when children experience safety in relationships and in their physical environments, they have a strong foundation for building resilience. Research consistently shows that sensitive, attuned parenting helps build the brain architecture that leads to this all-important capacity to bounce back from life's challenges, large and small.

Now, in Part II, we'll apply those and other insights toward *solutions.* We begin with a process that lies at the heart of parenting: helping our children learn to trust themselves by first learning to trust others. This special way of being with our children bolsters development and future mental health by teaching them to trust in themselves and in the world. Let's examine how we can help our children grow resilience and flex with life's challenges by considering the story of how one family approached a child's challenging behaviors.

COPING WITH LIFE'S UNEXPECTED MOMENTS

Joel and Ava sought my help because they were concerned about their six-year-old daughter, Jackie, who routinely had negative reactions to seemingly ordinary events. At birthday parties, she had such difficulty controlling her feelings and actions that she would annoy children with abrupt comments or name-calling. On the playground, she would shove playmates to get on the swing set or take her turn on the slide. She always demanded to know plans in advance and had difficulty coping with any surprises.

What finally prompted her parents to contact me was a particularly difficult encounter with Jackie's grandmother. A stable presence in their lives, she visited the family every few weeks from her home a couple of hours away. Jackie and her brother, Terrence, loved spending time with her watching videos, singing, taking walks, or just giggling. Their grandmother had recently visited for a day, spending time with the children while their parents were out shopping. After the kids went to bed that night, she stayed to chat with Joel and Ava. The evening went so late that she decided to spend the night—which she hadn't done before.

The next morning when the grandmother appeared in the family room, Terrence greeted her with a smile and a hug. But Jackie seemed less pleased by the surprise, reacting by first hiding behind a sofa, then popping up. "Hi poopoo face!" she shouted at her grandmother. Surprised, the grandmother ignored the remark. She sat with Terrence, who snuggled into her and showed her a scratch on his hand. Not quite knowing how to react, Jackie hid behind a chair again, then suddenly jumped up, emitting a loud growling sound: "Grrrrr!"

"Oh good morning," the grandmother replied. "Are you a tiger?" Jackie growled again, then began poking her brother, triggering a massive tussle between the pair. Amidst the ensuing struggle, Grandma's eyeglasses were broken. Watching from the kitchen, Joel strongly reprimanded Jackie, telling her to mind her manners and apologize to her grandmother.

Her parents had watched such scenarios play out so often that they had developed a standard response: stay calm, keep everyone safe, then reprimand Jackie, trying to teach her to improve her behavior. Clearly, though, their goodwill was wearing thin. More and more, the parents became so frustrated with Jackie's behavior that they shouted at her or sent her to her room. They were desperate to find ways to get through to Jackie and help their daughter find better ways to cope with the unexpected (and expected) experiences of her life.

SELF-REGULATION DEVELOPS FROM CONNECTION

The truth is, we all learn at different rates how to be resilient and manage life's shifting demands and challenges—to *regulate* our emotions and behaviors. *Self-regulation* is what enables us to respond to life's twists and turns with flexibility and forethought rather than exploding or acting impulsively.

Researchers describe self-regulation as the intentional control (regulation) of one's thoughts, emotions, and behaviors. In short, it's our ability to manage how we act and feel. Studies show that children who are self-regulated do better academically and socially. Not surprisingly, children who can control their emotions and behaviors have a head start on the playground and in the classroom. Parenting these children is easier, too. A child who is self-regulated can wait a few minutes if dinner is late, concentrate on homework even if they want to go outside and play, or tolerate car rides. They can sit until the recess bell rings or hold on to a question for a time rather than blurting it out while the teacher is talking. A self-regulated child can use words to navigate a conflict with a peer on the playground rather than simply pushing or hitting to get their way. Self-regulation allows children to adjust to life's challenges by using their own internal resources rather than needing an adult to navigate or mediate difficulties for them or with them.

Jackie's impulsive outbursts reflected her difficulty self-regulating—a challenge that showed up in her unpredictable behaviors. Her parents were stymied, upset, and often embarrassed by her outbursts. They expected her to be able to handle life's simple demands: behaving herself at family gatherings or holding it together when Grandma appeared unexpectedly. Often, though, Jackie seemed unable to do so, instead blurting out hurtful words even though she recognized that doing so was wrong. It's not uncommon for parents to assume their child can self-regulate when the child is nowhere close to being able to do it yet. That disparity is known as the expectation gap.

THE EXPECTATION GAP

Many parents assume that children are—or should be—capable of doing things that their brains simply aren't yet ready to do. Although we start to gain control over our emotions and behaviors as toddlers, honing that ability is a long process that continues into early adulthood and is nurtured through relationships. *However, as we all know, self-regulation isn't a simple developmental milestone that we reach at a certain age.* If it were, *we* would never "lose it" and fly off the handle as adults. As parents, we can help our children develop their own ability to control the way they act on their feelings. Fortunately, we can learn how to nurture the kinds of interactions that, over time, help our children develop better self-regulation.

Since Jackie was so verbal and conversant, her parents assumed that she had developed the ability to self-regulate. But Jackie continued to struggle: Her brother, Terrence, two years younger, often had better self-regulation skills than she did. As we began working together, I explained to her parents that Jackie's outbursts didn't stem from insufficient parenting, inconsistent discipline, or a lack of affection. They weren't raising a rude child. *Rather, though she was six years old, Jackie's ability to self-regulate was still under construction.* Her disruptive behaviors, such as the tussle with Terrence and Grandma, signaled that she was still developing the ability to control her emotions and behaviors.

I also told them there was a solution. We would start by strengthening Jackie's regulation and control by working on the *precursor* to the outbursts. The most important tool in helping her, I explained, would be **co-regulation**. *In short, we help our children learn to manage their emotions and behaviors through our loving interactions with them—through our relationships.*

SELF-REGULATION DEVELOPS THROUGH CO-REGULATION

Researchers have consistently found that co-regulation is the "superfood" that nourishes children's growing capacity to self-regulate. When we co-regulate with children, we help them to feel safe, and to tolerate and make sense of their sensations and basic feelings. It's helpful to think of co-regulating as sharing a connection. We co-regulate through our *emotional tone*, which is reflected in how we talk to and interact with our children. *When we do this, we accomplish a remarkable feat: We help regulate their body budget through our interactions.*

The experience of co-regulation starts when our children are infants, as we notice a baby's physical needs and respond to them, so that the baby feels better. The key for optimum co-regulation is that we respond to the baby's needs as they are happening. It's not helpful to feed a newborn an hour after the baby cries from hunger. We meet needs as they occur. We use our *responsive* interactions to figure out what they need in order for them to feel safe, calm, and comfortable. According to the psychologist Stuart Shanker, "a shared state of calmness results when we reframe another's behaviour and identify and reduce their stresses."

Mothers can support co-regulation even *before* the baby is born by maintaining a healthy lifestyle, getting adequate sleep and nutrition, receiving sufficient prenatal care, and minimizing stress during pregnancy. The prenatal environment can impact the baby's self-regulatory abilities after birth. Babies whose mothers have experienced intense stress or trauma may be more vulnerable physiologically, for example. Co-regulation helps a child feel understood, seen, and valued. It validates the child's emerging sense of self, making the child understand: *I'm seen, I matter, and my feelings matter to someone else.*

Some Examples of Successful Co-Regulation in Childhood:

- *Your* newborn *begins to cry, and you realize it's feeding time (again); you gently pick them up, nurse them, and they stop crying and comfortably gaze into your eyes.*

- *Your* **nine-month-old** *presses a button on a new toy that makes a surprising sound, and he immediately looks toward you with wide eyes and a scared look. You look at your child and say, "Oh my! That was a surprise!" with warmth and inflection, and he settles, continuing to explore the toy.*

- *On your* **toddler's** *first day of preschool, he stops midway between the parking lot and the school's front door to protest, saying that he doesn't want to go inside. You kneel down, tenderly acknowledging this important moment, letting him know with a soft and gentle voice that it's a big day and you are glad you are walking in together. He gazes at you, then takes your hand to walk toward the classroom.*

- *Your* **ten-year-old** *comes home from school and lets you know she's having difficulties with her friend group. You see a look of sadness on her face, thank her for letting you know, and invite her to tell you more. Her face softens and she leans into your body for a hug.*

- *You come home from a tough day at work. Your spouse or partner promptly gives you a loving hug, asking what you need: some dinner, some quiet time, a warm shower? Your body relaxes into the warm feeling of co-regulation, of being seen and attuned with. (Yes, co-regulation is a human process, beneficial not just for children but for parents, too!)*

In each of these examples, the power came from the caring and warm emotional tone. What's initially most important isn't necessarily what we say but, in a broader sense, how we *are* with our children. Of course, some encouraging words are welcomed as well, but as we learned, our brain detects a person's emotional tone milliseconds before it decodes language. *Indeed, the first step in helping a child who is struggling isn't telling, teaching, or giving instructions. It's being present with the child.*

Co-regulation is a cornerstone concept in the field of infant mental health and among scientists studying early development. It's less known in the general culture, however. Your child's pediatrician probably hasn't ever mentioned co-regulation at a routine checkup, nor has a teacher brought it up. *Yet, as a psychologist, I'm convinced that it's the single most important ingredient to help build our children's mental health and resilience.*

Co-regulation is what builds a child's future ability to manage life's ongoing challenges flexibly, face adversity, and form loving attachments with others. It also sets up the powerful modeling of empathy and caring for others. And it's a great way to make deposits into our child's body budget.

YOU DON'T HAVE TO BE PERFECT

As important as co-regulation is, it shouldn't add to parents' already considerable demands. Parents often feel judged or blamed, and I certainly don't want to add to your stress load or feelings of guilt. Certainly, our emotional tone matters, but remember that human beings are resilient, and research shows that raising young children naturally involves missing their cues *more often* than reading them accurately the first time. *Parenting is essentially a guessing game.* You don't have to "get" co-regulation perfectly. Imperfect parents (which we all are) can (and do) raise perfectly healthy children.

Evidence for that comes from Dr. Ed Tronick, a pioneering researcher on infant development, who has spent decades studying interactions between infants and their mothers. (His term *mutual regulation*, like *co-regulation*, describes how babies and their parents mutually affect each other's emotions and behaviors.) Dr. Tronick's research shows that when mothers try to figure out what their baby needs, they *rarely get it right on the first pass*. We don't magically understand what babies need from us; we need their help to figure it out. Those expected misses and mismatches are the norm. Only about 30 percent of mother-baby interactions are well matched or coordinated on the first pass, with the mother immediately and accurately detecting what the baby needs, says Dr. Tronick.

In addition, many mothers struggle with their own mental health after giving birth, some experiencing guilt and shame for having negative feelings or thoughts at a time when they are expected to be joyous. In the months after having my first child, with prolonged sleep deprivation and the stress of having to monitor the weight and progress of my premature baby, I experienced high levels of anxiety and dread and wondered what was wrong with me. Those feelings have sometimes recurred during stressful times, making me wonder if my own struggles were somehow damaging my children. We can expect a variety of feelings and emotions in reaction to parenthood, influenced by how we ourselves were parented, by our shifting bodily states, by hormone levels and sleep changes. If unpleasant feelings or thoughts become intense or overwhelming, however, it's important to check in with your health-care provider. Postpartum or ongoing intense anxiety and depression are medical conditions that require nurturant support and treatment.

Whether or not we face these challenges, discerning how to co-regulate with our babies and children is an ongoing learning process. Over time, we learn what our babies' behaviors mean, and we continue learning with them as they grow. As Dr. Tronick has written, "messiness is an inherent quality of infant–caregiver interactions, and therefore the task of creating shared meanings is a daunting one for infants, children, and adults alike." So if you occasionally—or frequently—miss the mark with your child, you are hardly alone. The research shows that learning and growth also happen in the *repair process*, when we try again and address the mismatch between what the child needs and what we *thought* the child needed *or what we were able to provide in the moment*.

We should realize that it's impossible to provide for a child what they need at every moment. Our pathway colors shift with life's demands, as do our children's. During the COVID-19 pandemic, many of us became all too familiar with the feeling of "hitting the wall" and landing in the red or blue pathway. I spoke to many parents who, during lockdowns, felt stretched to the limits and blamed themselves for their children's emotional and behavioral challenges. *The reality was that parents and children alike were behaving in ways that were consistent with having fewer internal resources and overdrawn body budgets.* It takes time to rebuild those resources. For now, if you are blaming yourself for negative interactions you've had with your child, know that children take in the aggregate of all our interactions. When we show up and repair as necessary, we help children develop confidence in the world and in themselves.

As parents, we don't always read our children accurately—and we shouldn't expect to. We help our children develop a healthy sense of themselves every time we repair our mistakes and nurture *ourselves* into feeling better when we are overwhelmed by the task of parenthood. Compassion and hope live in that repair process, and over time we can aim to make the positive interactions and experiences outweigh the negative ones. Along the way, when those negative experiences happen, we have the opportunity to model for our children that we can grow from them.

BOTTOM LINE: We don't have to be perfect to raise healthy kids. *Mismatches will always occur, but there will also always be the opportunity to repair—and that's where the growth comes from.* What's important is to learn from your child. Repairs are also opportunities to model mental flexibility for children. Admitting that you made a mistake is a powerful way to help children see us embrace our natural vulnerability as humans.

Some Examples of Mismatches and Repairs:

- *Your* nine-month-old, *eating in her high chair, begins to fuss. Thinking she's still hungry, you give her more food, but she pushes it away. (Mismatch) "Oh, you're all done?" you ask. She smiles and reaches out to be picked up. (Repair)*

- *Your* three-year-old *is enjoying playing with his cousins when you realize you're late to pick up your first-grader from school. You quickly grab the child's hand, saying, "It's time to go." He delays, and you lose it, saying, "You always make us late!" Startled, he begins to cry. (Mismatch) You slow down and sit with him on a step. "Oh my goodness. I was in a hurry, and I rushed so fast. Let's sit for a second, sweetheart. You wanted to play with your cousins, and you didn't make us late—that was on* me. *I lost track of time and I'm sorry I yelled at you." (Repair)*

- *Your* seven-year-old *brings in a lizard she found in the backyard, proudly announcing the new pet and setting it free in the kitchen. When you shriek and scold her for bringing it in, and demand she take it outside, she gets a tearful, dejected look on her face. (Mismatch) You apologize for reacting so quickly, telling her you're a bit scared of lizards, but you see how excited she is. (Repair) You settle on a mutually agreeable plan to release the creature into the backyard.*

When we allow our children to see that there is an alternative explanation for what we said, they are less likely to absorb a negative message they tell themselves (e.g., *I am selfish. I'm weak. Something's wrong with me.*). I understand that reparative work can bring up some tough feelings, but the good news is that the more you get it wrong as a parent, the more opportunities you have to set things right. *Children can grow from the reparative process, as they learn from watching your self-awareness and emotional flexibility.*

FACING CHALLENGES LEADS TO HARDINESS

Children develop hardiness and grit when they experience difficulties and, with the help of loving parents and other adults, work through those experiences. None of us develop hardiness without experiencing tolerable levels of discomfort or suffering. *Since we cannot protect our children from difficult experiences, it's comforting to know that hardiness is a possible benefit of facing challenges.* Children learn and grow best from their hardships through their parents' loving support. Indeed, this is how we all build the skills required to manage challenges. However, when children experience *only* the mismatches and ruptures, and not the repairs, stress accumulates, often exacting a serious toll. The psychiatrist Bruce Perry, who has studied how to help people recover from chronic stress, says that stress becomes toxic or traumatic when the child doesn't have adults in the picture who can offer support, which buffers it.

CO-REGULATION DOESN'T MEAN PAVING THE WAY: REMEMBER THE "JUST-RIGHT CHALLENGE"

In Chapter 1, we discussed the importance of using the right dose of support at the right time for each child. That doesn't mean we should aim to protect our children from all stressors (an impossible task, anyway). *Stress can be detrimental, but it can also be beneficial; and without tolerable levels of stress, we wouldn't have the opportunity to flex when faced with inevitable challenges.* In the words of Dr. Perry, "if moderate, predictable and patterned, it is stress that makes a system stronger and more functionally capable." Children build strength by handling a certain level of stress and, in the process, develop new skills to tolerate changes and a wider range of emotions.

Earlier I introduced the idea of the "just-right challenge." We shouldn't automatically do things for children that they can do for themselves. Rather, we should assess what they *can* do on their own and offer support when they need it—when we feel they have stretched far enough. This is another aspect of respectful co-regulation. *It's also important to give children plenty of opportunities for exploration on their own, with the parent in the background if help is needed.*

Without a degree of challenge we can't develop new skills—whether it's a basic skill such as matching shapes or an emotional skill like perseverance. If you rushed in the moment your child got frustrated trying to find the right shape to put in the shape sorter, they'd never learn to do it themselves. If you immediately tell your child what to say to a sibling they're struggling with, they won't have the opportunity to solve problems on their own, in the process practicing and seeing what works and doesn't work.

Here are some examples of parents meeting their child in the challenge zone successfully:

- *You arrive with your* four-year-old *to drop her off with a trusted babysitter, who has recently moved to a new house. In a hesitant voice, your teary-eyed child says she doesn't want to stay, but your intuition tells you this is a just-right challenge, so you walk her in and talk to the babysitter while admiring her new home. Your child softens her grip from your hand, you don't linger very long, and you say goodbye confidently. Fifteen minutes later, the babysitter texts you that they are happily drawing together.*

- *It's your* nine-year-old's *turn to share about their family at school. Together you've created a poster board complete with drawings and photos of siblings, grandparents, and the family dog. When your child tells you that they don't want to go to school on their sharing day, you lovingly tell them about the time you were scared about giving an important talk at work, normalizing and soothing their presentation jitters. You continue with the morning routine, letting them know that it's not an option to stay home. Returning from school later that day with a smile, they tell you with pride how much their peers enjoyed the presentation and how many questions they had.*

HOW MUCH DISTRESS IS TOO MUCH?

In each example, intuition led the way to make decisions that helped a child stretch into a manageable challenge. Of course, it's not always easy to know how to make complex parenting decisions. But using your child's nervous system as a guide makes the decisions easier. We can gauge the level of intensity of distress our child experiences (remember the level of distress continuum we discussed earlier) by thinking of it on a scale of 1 to 5. Each of these parents remained sturdy while their kids struggled with distress levels in the green pathway, or 1–3 range. If the distress levels were in the red pathway, or 4–5 range, then the children would have needed more co-regulation. We analyze the *frequency, intensity, and duration* of the child's actions and behaviors in order to discern whether a child can self-regulate, or if the child needs our help to do so, titrating the amount of support the child needs by observing their subjective level of distress.

In the case of a toddler learning to walk, a parent offers *just enough support*, stepping back so that the child can take their first few independent steps on their own. If the child hesitates, you offer encouragement with kind and supportive words, recognizing that the child is contemplating those first steps but is naturally hesitant. Their stress might be at a 1–2 level. We wait, and after a while, we might offer a hand or something to hold on to if we feel the toddler needs it. In short, we're encouraging the child to work within their challenge zone. Later in the day, that same toddler, now getting sleepy, might trip on a toy when trying to walk and cry very loudly, their body exhibiting a level of increased distress in the 4–5 range. That cry signifies a more immediate need for loving co-regulation, soothing, and comfort (and the realization that it's time to start the bedtime routine).

Sometimes, we aren't aware of the crucial role co-regulation plays in helping children manage stress and make sense of challenges. When I was a child, for example, my mother often worried excessively about the safety of her loved ones. My parents were both immigrants who worked long hours to support our family of seven. My mom had overcome many challenges in her own childhood, which predisposed her to this sense of concern. I still remember the anxious expression on her face on the frequent occasions that my father was late returning from work. Over time, I picked up on the worry. I would sit looking out the window, my heart beating fast as I eagerly awaited the sight of his van's headlights. I couldn't calm down until he walked through the door.

Of course, my mother had no idea that I was suffering. I was the oldest of four, and she had plenty do. Had she known, she certainly would have found a way to soothe my fears. In those moments, I needed co-regulation from my mom to soothe me from the emotional distress brought on by those frequent and intense experiences.

MIRROR NEURONS, MIRRORING, AND THE PARENT'S PLATFORM

It's natural and instinctive for parents to co-regulate with our babies and children. One reason: *mirror neurons*. In 1992, Italian researchers discovered cells in the human brain that help us understand the actions and experiences of others. That's why when we see our child suffering, for example, we are emotionally moved by the experience. Long before we knew of mirror neurons, the psychoanalyst Donald Winnicott described how a mother's face naturally "mirrors" her baby's feelings and needs—and recognized how important that was to the baby's growing sense of validation of their own experiences. *A child seeing and feeling an adult mirroring back emotions is itself a powerful form of co-regulation.*

In addition, mothers tend to meet our children's needs automatically and instinctually. Researchers have observed that when a mother witnesses her child's distress, she experiences changes in her *own* nervous system. Nearly every parent knows this experience of "feeling" your child's emotions in your own body. I have a strong sense of my internal feelings, so when my children were young and one of them was ill, I sometimes felt sick, too. If you've watched your child play a team sport, you might have felt as if you, too, were out there on the soccer field or basketball court. Our instincts, our mirror neurons, and our platforms all shift when our child is in distress. I often say that I feel like I carry four hearts in my body: mine and my three children's. The visceral experience of parenting is real.

Why is understanding and meeting our child's emotional needs through co-regulation so important? *When adults meet a child's physical and emotional needs from birth, we are building in self-regulation skills that will serve children through their life and into their adulthood.*

HOW YOU ARE IS MORE IMPORTANT THAN WHAT YOU SAY

I was blessed to experience co-regulation with my paternal grandmother, who lived in the Netherlands but spent most summers with my family. A quiet and anxious child, I eagerly anticipated her visits. My grandmother had long, flowing white hair that she kept in a bun. She smelled like lavender perfume and Dutch chocolate. She would bring European board games to play with us for rewards: special candies she brought in decorated tins. We took long walks, sometimes stopping to rest at a special tree. We talked and laughed from morning to night. Her sole purpose during these summers was to spend time with my siblings and me, four of her six grandchildren. She delighted in us as much as we delighted in her.

I'll never forget the night when a forest fire was raging in the nearby mountains a few miles from our home. I sat on my bed with my grandmother, watching bright red flames light up the night sky. I felt deeply afraid at first, but as she sat there, holding my hand and watching with me, I began to feel safe. We sat together for hours, marveling at the various colors of the flames as they edged up the mountains and over the ridge, away from our home. I even remember the yellow-and-gray-flowered dress she had on—the memory of co-regulation and feeling safe with my beloved grandmother is vivid, all these years later.

During some of the most difficult and stressful moments of my life, I have called on the memory of my grandmother, and her face and voice come to my mind clearly, even decades after she passed away. Picturing her never fails to help me feel centered and calmer. She helped me develop self-regulation, the ability to control my emotions and behaviors through my own actions. What a gift my grandmother gave to me: memories of co-regulation that wired my brain for eventual self-regulation.

When we co-regulate or share connection with our children, over time, the experiences form memories of feeling safe with another, and our children grow up expecting that others will meet their needs. *When you give your child loving experiences of connection, you are giving them the best possible head start in facing life's challenges.* That's because our nervous system remembers the safety of those experiences—both memories that we can recall and, even more, those that we can't remember. This sense of safety was my grandmother's gift to me.

Consider how boats stay stable in the water. When we meet our child's physical and emotional needs, we're playing the role a keel plays on a boat. The keel helps a boat stay upright and prevents it from being blown onto its side. If the keel is solid, then no matter what happens, the boat is stable. Unless the boat completely floods, it can weather storms and high winds without tipping over. Think of co-regulation as the keel: Whatever storms arise in the child's life, large or small, it helps the child maintain stability. When a child's upset encounters our calmness, we are helping to build their self-regulation.

Kids like Jackie have invisible needs that can land them in the red pathway. What their parents see are unpredictable behaviors. *Some children take longer to build their keels*—for various reasons we'll discuss in Chapter 6. So it's important to recognize those times when a child's behavior signals that the child needs co-regulation first and teaching later. If a child's keel isn't built yet, the child might require additional experiences of shared connection to help them feel more and more comfortable facing increasingly expanding challenges. *That was what Jackie needed, and what many children need when their behaviors continue to perplex their parents.*

CO-REGULATION: BUILDING LOVE AND TRUST

For that reason, when I worked with Jackie's family, our first priority wasn't to change Jackie's behavior; it was to better understand how she experienced the world. We needed to strengthen the shared connection that would help her tolerate and develop a bigger toolbox when she faced challenges to her platform. We discussed the expectation gap. Her parents were probably expecting more emotional regulation from her than she could sometimes muster, so they were often frustrated or disappointed—as was their daughter. While it was tempting for her parents to sometimes feel that their daughter was simply choosing to behave badly, it helped them to hear that she likely wanted to please them but that her platform wasn't in a place where she was able to do what was requested. Jackie wanted to please her parents, but in those moments when she could not control herself without the benefit of co-regulation, she just couldn't. So we took a different approach. Instead of trying to change her behaviors with rewards or consequences, we focused on strengthening her *self-regulation through the shared connection of co-regulation.*

IS CO-REGULATION SPOILING?

Parents often wonder whether practicing co-regulation is just another way of coddling or spoiling. I'm not suggesting that parents be permissive when the child actually needs limit-setting. Actually, coddling and co-regulating are quite different. One is incessantly indulging a child, never saying no. The other is attending to the child's physical and emotional needs. *Co-regulating doesn't mean that we make sure our child is always happy, or avoid having the child face any challenges.* Meeting a child's needs for safety and security doesn't mean giving the child everything they want or not letting the child struggle with difficult emotions. Indeed, growth happens when we give children the time and space to work things out.

The key is to observe and do your best to view the child's emotional reactions through the lens of their own experience of the world. Co-regulating doesn't mean abdicating our authority as parents but rather shifting our priorities. *Instead of deeming the child's behaviors as good or bad and then managing the behaviors, we examine the behaviors to find meaning and information about the child's needs, and pay attention to the significance of the stress held in a child's body.* We do this by observing their platform and pathway colors.

Co-regulating provides the most solid base for us to support children while they learn and grow, try new things, and sometimes struggle. Being able to regulate our *own* emotions helps us tolerate uncomfortable feelings so that we can help our children develop new skills and eventually manage distress on their own. If our children's behaviors make us feel anxious or stressed, that can undermine our role in co-regulation. This may feel like a burden, but don't worry. In the next chapter, I will share a plan to help alleviate the emotional load of parenting.

A parent can compassionately witness a child's struggle *and also* set firm and appropriate limits. In fact, doing so helps children develop a tolerance of a wide range of emotions—negative and positive.

We all have strong opinions about what behaviors mean—and how to help children control them. *Understanding co-regulation, we see that many challenging behaviors are simply reflections of a child's stressed-out nervous system adapting as well as it can.* As such, even the most challenging behaviors can be seen as protective and meaningful.

When Jackie appeared to be rude to her grandmother, her safety-detection system registered threat (despite the objectively safe circumstances), and the result was her red-pathway "rude" behavior. Behaviors tell us so much about what we can't see and what our children truly need from us in the form of co-regulation.

Co-regulating requires us to manage and tolerate our reactions to our children's wide range of different emotions so that we can stay calm when they need us to. That's often not easy.

It can be stressful when your child frequently has difficulty managing emotions or behaviors, but I encourage you not to give up hope. There's always something you can do to support your child and help them grow out of challenging phases. It is essential that your child experiences sufficient shared connection through easy and hard times with you and other trusted adults. *When difficult moments or conflicts occur, work through them together and implement repairs that work well for both of you. It's never too late to repair or to co-regulate.*

This is the heart of brain-body parenting: *When we respect children's subjective experiences based on their own body's reactions, we can titrate and customize our interactions.* Rather than simply rewarding "good" behaviors or taking advice from generic parenting guidelines, use your knowledge of your child's behaviors and signals from their body as a guide to help them expand their range of tolerating new challenges. With that understanding of the important role of co-regulation, let's discuss how to integrate this concept into your relationship with your child.

SERVE AND RETURN: BUILDING RESILIENCE THROUGH INTERACTIONS

Co-regulation involves back-and-forth interactions—a rhythm of shared connection and warmth. As in a tennis game, one person serves, the other returns, and the ball goes back and forth. Similarly, researchers and development specialists describe these as "serve-and-return" interactions. What one person does influences what the other does next. In a tennis game, a player decides how to return the serve based on the other player's anticipated actions. You alter your response based on where you anticipate the ball will land on your side of the net. Similarly, with our children, we alter our responses based on what they are serving to us. When such an exchange feels safe and nurturing, we have achieved co-regulation. When we can't figure out the kind of return or serve the other person needs, it's frustrating, and we can feel stuck or ineffective. That, too, is part of being a parent. It's not always easy to figure out how to calm and connect to a child. So don't judge yourself if it's difficult at first or as your child goes through various developmental stages or experiences.

The serve-and-return allows us to communicate with our children. It starts very early, as we fall in love with our babies through back-and-forth gazes and smiles. "The energy of reciprocity is one of sending care back and forth," says therapist Deb Dana. In other words, parenting isn't truly care*giving* but care-*sharing*. The communication is bidirectional. These early exchanges are the building blocks of communication, moments when a child learns to take turns with another person. At its core, human communication is a back-and-forth rhythm that need not involve words; facial expressions or gestures are often enough. This communication is foundational and beneficial, and, when it's working, it simply feels so good.

The serve-and-return of co-regulation can be as simple as a smile or as strong as rough-and-tumble play, as long as there's a back-and-forth and it's pleasurable.

A few things to consider if you feel something is missing in your shared connection with your child:

- *Does your* infant *look toward your face, even for moments at a time? (If not, do not worry; this behavior can take a bit of time to emerge, especially if your baby is premature.) A glance, gaze, or look from your newborn is a serve. It's remarkable that babies are hardwired to share connection with us with their eyes, even when they lack control over other body movements.*

- *Is your* nine-month-old *responding to you when you initiate a serve? Are they doing something in return—smiling, making a sound, reaching out with their arms, touching your hands, or putting the toys you offer into their mouth while looking at you? Is your baby initiating serves to you by looking or smiling, reaching their arms out to be picked up, or otherwise communicating with you?*

- *Does your* toddler *try to share their world with you? Do they point to things they desire or want to show you? Do they use their words or gestures to show, ask, or tell you something?*

- *Does your* school-age child *initiate conversations with you, play with you, or enjoy doing something together that involves back-and-forth interactions? Do they enjoy talking to you and telling you things, or simply being together while taking a walk, having meals with you, asking for help with homework, or otherwise hanging out?*

If your answer to these questions is "no" or "often no," it's useful to reflect on these basic communication building blocks and see if you can add pleasurable activities that encourage back-and-forth interactions. Sometimes our children need a little added joy to boost their serve-and-returns with us. If you find that the serve-and-returns I mentioned above for different ages are *lacking or completely absent*, then you may want to consult your pediatrician or developmental specialist for additional support. These positive encounters will lay the foundation for later interactions, such as when your child seeks out your advice when facing challenges or dilemmas and counts on you as a trusted adviser later when approaching teen and adult years.

Joel and Ava faced a parenting dilemma in the stark differences between Jackie and her brother, Terrence. Each child had different triggers and distinctive needs for shared connection. Grandma's unexpected overnight was a "serve" that presented a challenge for Jackie but was easy and pleasurable for Terrence. On the morning of the incident, Terrence joyfully returned Grandma's serve immediately, but seeing her grandmother unexpectedly sent Jackie into distress. Each child is different, and from everyday interactions we can learn a great deal about what a child needs.

With this understanding of the role of serve-and-returns in co-regulation, let's examine a four-part process to help you slow down and notice the *quality* of your serve-and-returns while also increasing shared connection and joy.

LOOK, OBSERVE, VALIDATE, AND EXPERIENCE

In thinking about co-regulation, it's helpful to remember the LOVE acronym, which makes this serve-and-return dynamic so powerful: look, observe, validate, and experience.

LOVE

LOOK: *Look at your child with "soft" eyes.* Soft eyes means that we widen our field of vision—literally and figuratively, which helps us have an open mind, free of judgment. By softening our eyes, we soften our hearts, opening to all we can learn. Having soft eyes helps you move toward respecting what your child's behaviors are telling you in this moment. The bonus: When we look with soft eyes, we communicate a message of acceptance, warmth, and affection.

OBSERVE: *Observe without judgment.* We are so quick to automatically judge our child's behaviors as good or bad, but as we have learned, behaviors are outward manifestations of a child's platform. Notice your child's face, gestures, and body, and be open to the idea that you have something important to learn from observing. Also observe your child's level of calmness or agitation. We want to pay attention to the myriad of information that observation brings. Watch what your child is doing during easy *and* difficult moments with an open sense of curiosity. When we observe without judgment, we appreciate that a child's behaviors are telling us something valuable, and we are open to learning what that is. When we give our child (and ourselves) the benefit of the doubt, we let go of preconceived ideas about what a child's behaviors mean and carry less self-judgment and self-blame for our role in contributing to their behaviors. Mantra: We're all doing the best we can with the information—and bodies—we have.

VALIDATE: *Validate your child's experience when you witness struggles.* If your child is struggling, then calmly try a serve-and-return that's soothing and connecting rather than judgmental or evaluative. Remember that difficult behaviors are your child's *nervous system's way* of asking for connection, for cues of safety from you. Your child wants to be seen and not feel alone. A very powerful form of validation is to simply bear witness to your child's struggles without automatically trying to solve them. Sometimes, that is enough, and the simple act of presence without judgment helps the child to regulate.

EXPERIENCE: *Experience safety together by sharing your green pathway with your child through your serve-and-returns.* Try various gentle serve-and-returns with your child, knowing that you probably won't get it right the first time. That's okay. We exercise our child's resilience muscles by helping them calm through our interactions when they need a deposit into their body budget. Try to move toward mutually enjoyable serve-and-return exchanges, even as you ask your child to tolerate and stretch through uncomfortable new experiences. In Chapter 6, we'll see how we can use insight into children's sensory preferences to further customize our interactions. This will enable you to increase the pleasure and joy of shared experiences, in turn helping your child stretch, develop new strengths, and increase their tolerance of new challenges.

For Joel and Ava, the LOVE exercise helped to guide our work in *strengthening* Jackie's platform, by enhancing her capacity for self-regulation through the process of co-regulation and shared connection. Previously, they had focused on simply changing her behaviors through consequences or time-outs. Now, they used the check-in, described in the last chapter. They had a new lens through which to see and understand her behaviors as reflecting her platform and body budget. They validated her—acknowledging, to themselves, how difficult self-regulation was for her and connecting with her in the green pathway.

Joel and Ava also came to understand that Jackie's nervous system often registered everyday events as threatening, and that tendency came with a heavy cost to her and her family. So her behaviors were actually *protective* to her platform. Previously, they often ignored her "inappropriate" behaviors or gave her consequences for them because they viewed them as negative and obnoxious. They changed their approach from their go-to strategies—reprimanding, consequencing, incentivizing—to instead making it a priority to co-regulate.

Previously, when Jackie would have tantrums over seemingly small disappointments—such as getting the wrong color sprinkles on her ice-cream cone—her parents would try to help her see the bright side. If they were stressed themselves, they'd suggest that she was overreacting. When, instead, they looked at her with soft eyes, they saw that she wasn't being rude intentionally. *She hadn't fully developed her mental flexibility or the ability to shift and adapt quickly when something unexpected happens, a capacity that can require many years to develop.* With this new understanding of their expectation gap, they shifted from automatically reminding her to behave better to first finding patience in the green pathway and appealing to her in a gentle tone: "I can see this is hard for you, Jackie." *They shifted simply to "being"—before teaching, correcting, or issuing consequences.*

With this shift, Joel and Ava felt more compassion because they saw their daughter's behaviors as by-products of her developing sense of self-control and a platform that required support through co-regulation. *Though many people assume that discipline is the best way to help children behave better, co-regulation is the key to developing self-regulation, which results in better behaviors as the natural end product.*

HOW UNDERSTANDING CO-REGULATION HELPED JACKIE AND HER FAMILY

As we worked together, Jackie's parents began to look at their child differently. Using their new skill of nonjudgmental observation, they discovered that their daughter was unusually sensitive to subtle changes in routines or in her environment. Digging deeper, we discovered that Jackie craved predictability because she was easily triggered into distress by a range of sensations she felt in her body. Many ordinary sounds in the environment—noises from the TV, the crying of her sibling or another child—triggered her safety-detection system. Looking with soft eyes and observing without judgment, her parents noticed that when Jackie encountered background and foreground noises simultaneously (like in a busy restaurant or mall), she began to bite her nails and rapidly repeated the same question—signals that her platform was moving from strong (the green pathway) to vulnerable (the red pathway).

This was enlightening for her parents, who had never considered that Jackie's *body (via her brain)* was registering threat and taking hits during objectively safe everyday experiences such as going to the park. They felt deep compassion when they realized that large family gatherings could send her body into distress. Knowing this, they shifted their "returns" from her "serves." *They now saw her "bad" behaviors as signaling distress rather than being rude.*

I explained that children with such sensitivities often benefit from routines and sameness—and may be perceived as inflexible. A child who feels—through no fault of their own—overwhelmed in their body craves predictability. But that's different from being *intentionally* bossy and rude. Still, what is adaptive to the child's system can be difficult on parents. So it's crucial to remember that many difficult behaviors aren't always deliberate or planned but are often responses to stress.

Jackie's need for routine and predictability explained why changes—even positive ones, like having her grandmother stay overnight—triggered Jackie into her challenging behaviors. Changes in routines threw her into distress, and she wasn't able to seek help or use words to describe what was happening—yet. Finally, her parents had an understanding of Jackie's behaviors that went beyond willfulness or deficiencies in their parenting. As my colleague Ross Greene, a psychologist, famously says: "Kids do well if they can." I agree. And when they can't, we should become curious and ask why. (The same goes for parents.)

As it turned out, their new understanding of shared connection as a parenting tool was transformative for Jackie and her family. Her parents came to realize that Jackie needed more shared connection and co-regulation in order to become more flexible and to feel safer in her body's responses to experiences—particularly when she encountered the unexpected.

Once her parents began to see that her behaviors emerged from an underdeveloped ability to self-regulate, they focused on co-regulation. At first, they simply made an effort to have more fun together, finding more moments of joyful connection. Her parents came to fear Jackie's unpredictable behaviors less. They *did* maintain consistent house rules and boundaries, calmly and authoritatively pointing out behaviors they deemed unacceptable. After all, they were the parents and it was a priority to instill their family's principles and values in Jackie and her brother. *But the focus turned from putting out fires to shoring up the co-regulation.*

When they observed subtle signs that Jackie was veering into the red pathway, they would draw closer to her emotionally and physically through affectionate gazes, making small talk, and enlisting her help with simple, fun chores. They also began to inquire gently about what she might be feeling inside her body. Most of the time she couldn't answer, but sometimes she was able to say "yucky" or "happy." This was the first step in getting to the source of her dysregulation: helping her *begin to notice her bodily sensations.* Whether or not she told them, *when they realized her body budget was depleting, they would engage in positive, platform-building serve-and-return interactions.* For example, when she seemed worried about going to a family gathering, they helped to prepare her by talking about who would be there and what it might look and sound like. This simple act helped her nervous system prepare for the required challenges ahead of time, helping her body manage the stressors of family gatherings more easily.

They used active co-regulation and shared connection through their interactions with her and appreciated how sensitive she could be at times. They put renewed effort into preparing her before the sorts of outings that had typically resulted in struggles in the past. They realized that many of the activities in Jackie's life were *above* her challenge zone, so they treated her with more empathy and worked on co-regulating to help her develop her self-regulation.

Knowing that she could be overly sensitive to sounds, I loaned her a pair of child-size headphones to wear at busy events. That gave her something to do when sounds felt overwhelming.

After a few months, Joel and Ava noticed that they—and Jackie—were more relaxed. As Joel and Ava started to feel better, so did Jackie. They became more present and shifted from perceiving their child as a behavior problem (and themselves as inadequate disciplinarians) to respecting their child's need to develop self-regulation through the shared connection of co-regulation. This shift was healing, and, in time, Jackie's behavioral challenges decreased and faded away. At one of my last meetings with her family, Jackie returned the headphones to me along with a drawing of a sun as a thank-you, shyly smiling and telling me she didn't need them anymore.

RESILIENCE-BUILDING TIP: Self-regulation is a process that begins at birth through a relationship of *co-regulation* with a loving adult who reads a child's cues in real time and provides them with years of connected, affirming experiences. Emotional co-regulation with caring adults leads to successful self-regulation.

A child's experience of co-regulation eventually leads to the ability to manage one's own emotions rather than being overwhelmed by them. It's a day-by-day process and a cornerstone that supports children's future emotional health and resilience.

Even under the best circumstances, parenting is often challenging and exhausting. Co-regulation can feel good, but it takes a lot of effort and energy. How can we be present and intentional for our children when we feel drained? Next, we'll consider how we can maintain energy and stamina, because *we* are the most important tool in the toolbox.

5

Taking Care of Yourself

Nurturing your own development isn't selfish. It's actually a great gift to other people.

—Rick Hanson, PhD

There is no magic formula for making sure our kids are happy and whole. But one factor can stack the odds in their favor: sturdy parents. When our platform is strong, we are better able to guide, teach, nurture, and set limits for our children. And of course we all have those less-than-stellar times when we lose it and make parenting choices we later regret. As important as it is to co-regulate with our children, we need to be okay ourselves. That doesn't mean we have to feel our best all the time, but rather we have to feel okay *enough* to share time with our children in the green pathway—where the really "good stuff" happens: the snuggles, the giggles, the moments of quiet tenderness and comfort, or sturdiness, when our child needs it.

WHERE ARE YOU LANDING IN YOUR CHILD'S NERVOUS SYSTEM?

As we've discussed, one of the ways we make deposits in our child's body budget is through co-regulation, a process that contributes to a child's ability to self-regulate. And the critical ingredient of co-regulation? Ourselves—our platform, which shows up in our tone of voice, our emotional expressions, our gestures, our pacing, and our words. In this chapter, we'll examine ways we can improve the chances that we show up as our most regulated selves, without judgment and with as much self-compassion as possible. It will offer what research shows are the most effective ways to make deposits into our own body budgets so that we have enough resources to support our children. But most of all, taking care of ourselves improves and maintains physical and mental health, so it's simply a good idea, regardless of whether or not you are a parent.

I was a child psychologist before my children were born, so I felt overly qualified to care for their emotional health. I was well acquainted with my field's parenting and attachment literature and truly thought that the emotional part of parenting would be a walk in the park! Oh, was I wrong.

WATCHING MYSELF AS A YOUNG MOM

Not long ago, I was watching some old family videos when one particular moment stopped me in my tracks. It showed my husband and me, in the yard playing with our three young children. Suddenly, the youngest started to shriek. Her older sister, a sensitive and quiet child who rarely shouted, started yelling back. Then I did something that made me cringe as I watched it all these years later: "You're out of line!" I said loudly and sternly. Then I sent my four-year-old (the one who had shouted back) to a time-out. Shocked, my daughter looked at me with wide eyes, clearly wondering what she had done wrong. Watching myself on the video, I saw a stressed-out, dysregulated mother who seemed most worried that the next-door neighbors would judge me for having screaming kids. And I saw a little girl who was asserting herself, appropriately, and my overreaction to her behavior.

Clearly, I had no idea where my actions were landing in my child. I used a hammer when all I needed was a feather—or maybe even nothing. I should have simply witnessed with curiosity rather than losing control of myself. I obviously felt so self-conscious that my daughter's screaming had sent me into the red path, and I projected my own insecurity onto her, judging her harshly and punishing her. In hindsight, however, I know that I was doing the best I could as a mom. In that moment, I was completely disconnected from my own regulation and my nervous system. Sure, I was a child psychologist, but at the time I knew nothing about platforms, safety systems, or the pathways of the nervous system. My field hadn't incorporated those concepts—and still hasn't.

One lesson: It's useful to know where our actions and words are landing in our children. And then to learn how we can be as self-regulated as possible. *Because we, and our own sense of well-being, are the most important tools in our parenting tool chest.*

Many parents, particularly mothers, have confided in me that they worry that their own burdens—depression, anxiety, or inadequate health care, mental-health support, or financial resources—have taken a toll on their child's development. Indeed, many parents around the world have no choice but to "pour from an empty cup" every day.

We attribute such importance to a child's first years that many loving and compassionate moms carry hidden guilt about their difficulty showing up for their babies or children. I know that I often wondered if my own anxiety and burdens were harming my children's development or making them anxious as well. To give perspective, let me offer this reassurance: Of course our life circumstances impact our parenting, but our child's (and our own) development is dynamic. I've seen how the brain and body protect and adapt to our life experiences, and the window of development never closes. I've helped eighty-five-year-olds change their thoughts and improve their outlook! So if you feel concerned about how you've interacted with your child in the past, please find compassion for yourself. And know that the power of *present* relationships changes the way we see the world in the *future. There's always the opportunity to create safety and security in our relationship with our child (and with ourselves). Each day brings new opportunities for connection, new ways to find compassion for our children and ourselves.*

We all know what it's like to be in a room with a stressed-out person. It also follows that what's good for you is also good for your child. Our children benefit on a biological (and emotional) level when we are in balance. This is one reason why self-care is so important.

That's the focus of this chapter: what parents can do to reduce our own stress. What matters most isn't what's generically good for mothers or fathers, but whatever brings you into connection with yourself—and adds to *your* body budget. One note: Since this is a parenting book, you might be tempted to skip over a chapter focused on taking care of yourself, but please don't. Because at the heart of parenting is our own well-being. Our platforms drive our parenting behaviors just as they drive our children's behaviors. This isn't about beating ourselves up if we don't go to the gym or practice yoga; it's about remembering that we need to nurture our bodies and brains in order to feel regulated—just like our kids.

PARENTAL WORRIES STICK LIKE VELCRO

One reason it's so stressful to be a parent is because we—like all human beings—tend to pay more attention to negative experiences than positive ones. This is the result of a function known as the negativity bias, a process by which the brain prioritizes negative experiences over positive ones. This survival-based instinct is an adaptive response—paying attention to threats in our environment gave our ancestors an advantage. If our brains and bodies didn't react to a thunderstorm or an oncoming herd of animals, we might not have lived to see another day.

Over time, this bias—paying attention to the negative things in our lives—can lay the groundwork for behavior patterns that don't always benefit us. The psychologist Rick Hanson explains the long-term effects of the negativity bias this way: Bad experiences stick to us like Velcro and good ones slide off like Teflon. I know that my own brain was sticky with fears about my children's well-being and I often couldn't relish the present moment because I was so busy worrying about them. Did they eat enough healthy food today? Was I giving them enough extracurricular activities? Were they overscheduled or underscheduled? Traditional psychology might label my tendency to worry as anxiety, but Hanson's view is more neutral, recognizing that the human brain has evolved to detect threat and protect our young. And it's not always about physical threats to our children. As Dr. Hanson reminds us, negative experiences are simply more "sticky." For example, many of us tend to focus on a negative comment from a colleague or supervisor, overriding the otherwise good news of the day. Perhaps understanding this natural tendency can help us be more compassionate toward ourselves.

Still, while prioritizing threats helped our ancestors survive, for modern parents it can be a burden. Understanding this tendency helped me move from seeing myself as a neurotic or anxious mother to viewing myself as a sensitive human who experiences the negativity bias, just like everybody else. Still, we can take measures to make friends with this bias, live with it, and offset it—as we'll see in the pages ahead.

Indeed, one of the most important parenting skills of all is getting to know and care for ourselves.

THE EVIDENCE SUPPORTING SELF-CARE

Several years ago I cochaired a conference on parental stress and how to build resilience in ourselves and our children. I was captivated by one of the keynote speakers, Dr. Elissa Epel, a luminescent woman who radiated calm and inner strength. Dr. Epel had worked with Dr. Elizabeth Blackburn, who won a Nobel Prize for her pioneering work on telomeres, parts of chromosomes that indicate how fast your cells are aging. Dr. Epel described how we could reduce stress and even reverse aging. I listened intently. After all, I was an overscheduled mother with a hardworking nervous system. I also witnessed stress and trauma in my work each day. Dr. Epel's studies found that simple self-care strategies such as mindfulness and meditation breaks helped parents like me slow down cellular aging.

Then there was the massive study two decades ago by the National Research Council and the Institute of Medicine. It integrated the science of early childhood development into a comprehensive summary. The result was *From Neurons to Neighborhoods: The Science of Early Childhood Development*, which translated science into practice. I considered its most important finding to be something my colleagues and I already knew: that nurturing relationships are essential to children's brain development and future resilience.

More recently, experts and researchers from the now renamed National Academies of Sciences, Engineering, and Medicine published another groundbreaking consensus study about early childhood: *Vibrant and Healthy Kids*. It reconfirmed the importance of early relationships but also included a new finding, even more groundbreaking than the first: "Ensuring the well-being of caregivers by supporting and caring for them is critical for healthy child development." *In other words, parents need nurturing, too.* And it's critical for our children's well-being.

While that may sound like common sense, child-development researchers have typically focused on what parents *do* to help their children develop, not how the parents *feel* or whether they are supported and nurtured. Yet many parents are overburdened and under–cared for. The burden is *exponentially higher* for those who struggle with financial or food/shelter insecurity, those who have inadequate health insurance, those who experience racism and implicit bias, or those who are single parents. Findings by the National Institute of Allergy and Infectious Diseases have shown that members of minority groups are more likely than others to suffer from chronic illnesses such as hypertension, lung disease, and diabetes, attributing this disparity to "social determinants of health dating back to disadvantageous conditions that some people of color find themselves in from birth regarding the availability of an adequate diet, access to health care, and the undeniable effects of racism in our society." Research validates parents' need for increased physical and emotional support, help, and recognition—a message I have shared with my patients for years.

Amelia and Silas

My patient Amelia, a single mother, started to cry when she told me about her difficult transition into motherhood. She lived with her baby, Silas, and her mother at her mother's home. Amelia worked at an insurance office and her mother stayed home with the baby. Amelia's employer's parental-leave policy gave her several months at home to nurture her newborn. But as much as she felt joyful and loving toward her child, when Silas would cry, Amelia often felt intense anxiety and distress.

Sitting in my office after her pediatrician referred her to me, Amelia explained that she understood that big emotions were an expected part of toddlerhood, yet Silas's crying routinely triggered a sudden and unfamiliar sense of dread. If the tears continued for more than ten minutes, her heart would race and she felt sweaty and anxious. Sometimes it was so difficult that she had to leave the house and take a walk around the block while her mother took over.

Amelia didn't know why her son's distress so triggered her. And when the pediatrician reassured her that Silas was healthy and on target in his development, it did little to ease her nerves. Yes, she felt gratitude and love for her son, but her transition to parenthood was uneasy, and she felt like the bottom could fall out at any time emotionally. Being the sole provider for the family added an additional source of stress and burden.

Tyrone and Dana

Tyrone and Dana contacted me for support during the COVID-19 pandemic. The couple were reeling after months of lockdown, during which they both transitioned to working at the home they shared with their active seven-year-old son, Jaheem. Pre-pandemic, Jaheem had been happily attending a school and after-care program he loved, while his parents worked full-time in jobs they enjoyed. As the months wore on, the parents found they were arguing more with each other and interacting more negatively with Jaheem than they could ever remember. Before COVID, their main concerns had been typical: Did Jaheem complete his homework? Was he getting enough outdoor and exercise time? Were they reading to him enough at night?

Now on top of those questions came added stress and the added guilt of having to set firm limits on Jaheem so his parents could focus on their own work. He would constantly pop in and ask them questions. Tyrone admitted that he often snapped at Jaheem for interrupting him, then felt badly for doing so. Finally, the stress compounded because the couple began to disagree about how to handle Tyrone's emerging and uncharacteristic defiant behaviors, like refusing to put his toys away or sit still for Zoom class sessions.

These two families' experiences are hardly unusual. Just as our children's emotions can be triggered by countless experiences, so can ours. The good news is that there are ways to cope. In the coming pages, we'll explore some powerful, research-based tools to help parents strengthen their emotional and physical health.

GETTING BACK TO THE GREEN PATHWAY: TOOLS FOR REDUCING PARENT STRESS

Just like our kids, when we take care of our body's needs, we have the platform to be more grounded emotionally. When we feel better, we're better able to co-regulate with our child.

Even though many of us know what good self-care looks like, so often it's hard to "just" exercise, get more sleep, or eat healthier meals, as we are often told and encouraged to do. The reality of parenting is that sometimes we sacrifice our own needs in order to support our children and put food on the table, pay the rent, and maintain stability. And, as Tyrone and Dana experienced, the global pandemic only increased the stress of preexisting parenting pressures.

If you're struggling to maintain your emotional regulation, I want you to answer the questions below, using the *average* of a week's tracking:

- *Are you getting seven to eight hours of sleep each night?*
- *Are you drinking six to eight glasses of water per day?*
- *Are you eating a variety of nutritious foods each day?*

- *Are you moving your body each day, either through formal exercise or through your daily activities?*

Chances are, you probably answered "no" to at least one of the questions, especially if you're the parent of a newborn, infant, or toddler. That's okay! The first step is to realize that something's missing. When you're overwhelmed, it's easy to forget to drink water or eat a proper meal or move your body from the chair where you spend the day working.

If you aren't getting enough sleep, staying hydrated, or eating well—or if you're too socially isolated—your physical health will likely reflect the consequences, sooner or later. Depending on your genetics and life experiences, you may not see the impact until you are older. But nobody is immune to the negative consequences of insufficient self-care, especially if the neglect is long-term.

Looking back, I wish I had given the idea of self-care some consideration when my children were young and I felt invincible. I was so busy that I always made my needs secondary to what I perceived as my children's needs. After all, I had plenty of energy, loved my work and busy life, and had never heard of a body budget. My education and training in mental health reflected the mind/body split: anxiety, for example, was treated as a mental concept, unrelated to the body. But as we have seen, the distinction between mind and body is a false one. Caring for your physical body *is* caring for your mental health. Self-care *is* mental-health care. It feeds our ability to co-regulate because it's impossible to parent well on a chronically depleted body budget. The mental energy, planning, and constancy of parenting are taxing on us physically and emotionally, and we must learn to prioritize our well-being.

Let's begin with the basics. Simply remembering to eat nutritious foods and hydrating yourself will help your body get the nutrients it needs. And moving your body as a part of the day or having a basic exercise plan is important to maintaining your health, especially in the long term. That said, the most important factor for parents' health isn't what we eat or how much we exercise but something we don't consider enough: sleep.

SLEEP IS YOUR LIFE SUPPORT SYSTEM

If you answered "no" to the question above about getting seven to eight hours of sleep a night, you are hardly alone, but you are also at risk—particularly if you suffer from sleep deprivation over months or years. Matt Walker, a leading sleep researcher and professor of neuroscience and psychology at the University of California, Berkeley, minces no words in his popular TED Talk, "Sleep Is Your Superpower." "Sleep is not an optional lifestyle luxury," he says. "Sleep is a non-negotiable biological necessity. It is your life support system." Sleep supports every system of our body, including our cardiovascular system, our neuroendocrine system, our immune system, and our ability to co-regulate, to think and make decisions in our occupations and in our parenting.

Yet every parent knows that once you have children, your sleep is never the same as it was before. I often say, only half in jest, that my last fully restful night's sleep was the night before my first child was born. After that, a part of my brain was always listening for crying, calling for me, a cough, or my sleepwalking child. Nights became a constant stream of sleep interruptions. There was always a precious someone—then two, then three—who might have needed me in the middle of the night.

To be sure, sleep deprivation isn't an easy problem to solve in our culture. That's partly because so many of us are geographically (or otherwise) separated from extended family and readily available caregivers. We don't live in extended family groups that were the norm in earlier eras, so most parents bear the load of child-rearing on their own.

I was raised in a multigenerational home with three adults: my parents and my maternal grandmother. Fortunately, one of the three could always stand in when another needed a quick break or nap, was sick, or had to run an errand or care for another child. Most families today don't have that luxury. Many of us are on an endless cycle of working and parenting, and sometimes the only time to get things done is when your children are asleep. *Still, sleep is our life-support system, and it's important not to let sleep deprivation become chronic.*

Until your baby grows and establishes a predictable sleep pattern, it's useful to be creative. (Chapter 7 will explore how to help children sleep.) You might find ways of sharing watch over your baby, or perhaps just reduce your expectations of what you can accomplish each day. Take turns with your spouse or partner handling nighttime wake-ups. Or just heed the sage wisdom for parents of newborns: "Nap when your baby naps." For parents of older children, you can model good sleep hygiene by having predictable bedtimes and healthy bedtime routines that include a time to wind down together, brush teeth, read a book or snuggle, and ease into sleep. We can show our children how we nurture sleep for ourselves and for them. In the short term, being more rested will help you be more patient and present with your child. And years down the road it could make the difference between having a medical condition or not. It's that important.

HUMAN CONNECTION: CAN YOU SPARE A FEW MINUTES A DAY?

Along with food, water, and sleep, there's another essential parenting nutrient: connection with other adults with whom you feel safe.

Like our children, we survive and thrive through human connection, the single most important factor in building resilience. Again, it's essential for our children's development that we feel cared for as parents—that we feel seen, affirmed, and loved, not just by our children but by other adults. Suniya Luthar, a prominent stress and parenting researcher, poses a question: "*Who mothers mommy?*"

When I heard Dr. Luthar ask that question at a psychology conference years ago, I smiled inside because it felt so affirming. Of course, for most mommies, the answer is . . . nobody. In many Western cultures, particularly the United States, we parent in isolation from extended families and neighborhood communities. Since my grandmother lived with us from the time my parents were married, my parents had a built-in (and capable and adoring) support system. They never had to parent on their own. I have no doubt that this contributed to their ability to open and run a successful business.

I chose to raise my own children more within our small nuclear family. I savored the privacy and the control I had over my kids' experiences, and the security I felt around that control. And frankly, I enjoyed co-parenting with my husband without much outside help or interference. Still, Dr. Luthar's question is important. Who mothers (and fathers) us? Who is there for you emotionally? Her extensive research emphasizes how important it is for mothers and fathers to have social and emotional support from friends and/or family.

That's because loneliness and social isolation pose risks to physical and mental well-being. A large study conducted by Cigna revealed that nearly half of the twenty thousand adults surveyed sometimes or always felt alone or left out. We need other humans to help us feel safe and secure. Think back to how you felt during the pandemic lockdowns. Almost every parent I worked with and spoke to during that time felt isolated, stretched, and/or anxious.

We underestimate how vital human connection is to our well-being until we are forced to live without it. The good news is, even short moments of connection with another caring adult can help reduce your stress and help you feel more grounded.

Consider a few questions that can help you assess your need for emotional support:

- *Do you have a co-parent, a spouse, a parent, friends, family, or others in your community with whom you feel safe, loved, and accepted?*
- *Can that person give and receive, so that you both experience being vulnerable and share concerns with each other without filtering or worrying about being judged?*
- *Is that person accessible to you when you need them or, even better, on a regular basis to spend time together?*

It's important for every parent to have a person who can witness your struggles, with whom you feel safe, and whom you trust—who can help you to feel seen and loved, essentially to co-regulate with you. Remember that it's not just babies or children who need co-regulation with other humans. We all do! *Try to have at least a couple of minutes of warm connection with one person each day, either virtually or in person.* This is especially important if you are a single parent. Call a friend or plan a time to take a walk or have a cup of coffee with someone you trust. We all benefit from these relationships.

If you don't have that kind of support at the moment, that's okay. Maybe that's too much for right now but something to consider for the future. These suggestions aren't meant to add burden but to make life easier.

SELF-CARE BEGINS WITH AWARENESS

Just as the power of observation is crucial to understanding our children's behaviors, we must also observe ourselves. Awareness is our portal to gaining compassion for our platform, how our body and brain are faring in the moment and over time. Remember the check-in from Chapter 3? When we have awareness, it allows us to appreciate, and bring to consciousness, how calm or agitated we feel. If we know we are in the red, that simple moment of awareness can be enough to prevent us from yelling at a child and instead choose something else to do that will be more affirming and relational. When we practice becoming more aware, it leads to feeling more grounded and balanced and provides a momentary pause that can prevent us from saying or doing things that are not in our child's—or our own—best interest.

When I was a girl and my beloved grandmother was pondering a problem, she would sigh, take a deep breath, and slowly repeat the word "so." She did this when she couldn't find the right word in English, her second language, or when something unexpected happened and she needed time to think.

The letters in "so" also happen to stand for the two steps to self-awareness: "**s**topping" and "**o**bserving." These two steps comprise Step 1 of the self-check-in below, which helps you identify which pathway you are on. I'd like to think my grandmother instinctively knew that slowing down and pausing was useful, decades before mindfulness became so popular.

SELF-CHECK-IN: Take a moment to **stop** and **observe** what's happening inside your body. Simply observe anything you are experiencing: a physical sensation such as the beating of your heart, a pain somewhere, thirst or hunger, an emotion, a thought. Observe without evaluating it as good or bad. Simply notice. If you are able to notice *anything*, congratulations! You've just had a moment of awareness, of mindfulness. If nothing's coming up or negative sensations or feelings do, that's fine, too; allow them in and then try to recognize the experience without judgment. There's no right or wrong here. Awareness of present experience can feel strange if you aren't used to it. Please don't be self-critical if it feels uncomfortable to slow and observe your mind or sensations; it's a new experience for many of us. *Awareness is a key ingredient of self-regulation for ourselves and our children.*

Some people find it difficult to stop or slow down enough to allow for awareness. If you find that the effort is creating stress in your body or mind, simply take note of that with compassion and without judgment. Some of us have engines that run fast as a protective or adaptive way of helping us stay in balance. That's okay. If slowing down sets off your safety sensor, you can honor that with curiosity. If you feel that slowing down makes you uncomfortable, simply stop and try it again at some point in the future when and if you feel ready.

Many parents tell me that they don't have *time* to notice their body's sensations. I understand. As a young mom, I was often so busy that I didn't even notice if I was thirsty, so some days I forgot to drink water. I was in constant motion and slowing down wasn't practical—until one day when I landed at my doctor's office for symptoms of dehydration. It proved a valuable lesson, making me realize how out of touch I was with many of my body's sensations—including thirst.

Self-observation can help you figure out what you need to fill up your tank and strengthen your nervous system for the herculean job of parenting. When I was actively raising three children, self-focus and self-care felt like luxuries and even seemed counter to my instincts, which surely were informed by cultural messaging that maternal sacrifice is a virtue. I preferred to spend my "free" time and energy on my children. I loved being a mother and I paid little notice to the toll that stress was taking on my life.

Today when I reminisce with my adult children about that time, we laugh and cringe at our memories, now shared with love and compassion. My husband and I were ambitious parents on a dual track of parenting and building careers. I often operated with little self-awareness. Once, calling a client from home, I grew so frustrated that my kids wouldn't be quiet that I threw a hairbrush in their direction (cringe). Another time, I drove to the shopping mall, not realizing until we arrived that I had forgotten to put on my shoes (laugh). Yes, my parenting life was sometimes an out-of-body experience. If only I had understood the importance of cultivating awareness and taking the time to meet the needs of my own nervous system, I would have been a calmer and healthier parent.

PRACTICE: One way to develop a greater awareness of your needs—and the pathway you are operating from at any given moment—is to practice mindfulness. Decades of study have demonstrated the positive impact of mindfulness on mental and physical health. You can start a mindfulness practice by simply taking a momentary pause several times a day to observe sensations or emotions. You can add a boost to your observation by setting aside a few minutes for moments of mindfulness—simply paying attention to the present moment without judgment. These brief periods not only help you become more connected to yourself but also reduce stress. There are many free apps available with guided mindfulness exercises, some as brief as one minute.

Whether we are grounded in our own calm state, or notice we've lost our sense of calm, the goal isn't to be a perfect parent but rather to navigate the inevitable twists and turns with awareness of our own bodily sensations, emotions, and thoughts. When we do, we can find comfort in allowing ourselves to feel what's there and to help ourselves rather than shame ourselves.

SELF-COMPASSION

In today's high-pressure culture, research shows that many parents either feel judged by others or judge themselves negatively much of the time. A large study found that nine out of ten parents feel judged almost all the time. So if you feel judged or judge yourself harshly, you're not alone. It seems that it's easier to have compassion for our children than it is to have compassion for ourselves.

When I was in the throes of parenting my young children, I might have deemed the concept of self-compassion nice but not necessary—perhaps even self-indulgent. But a large body of science has shown that self-compassion benefits our physical and mental health and overall well-being, and those benefits trickle down to our kids. Self-compassion is also a helpful tool when our platform shifts unexpectedly—as it inevitably does on the parenting journey. It allows us to do something soothing with our awareness, enabling us to become more compassionate and present with our children.

Dr. Kristin Neff, a pioneering researcher on self-compassion at the University of Texas at Austin, has done transformative research on the topic. She and a colleague, psychologist Dr. Chris Germer, developed a mindful self-compassion (MSC) training program to teach cutting-edge self-compassion practices. They teach mindful self-compassion: present-moment awareness combined with the action of sending kindness to oneself. Their pilot study revealed that participants in their course increased their levels of self-compassion and also felt less anxiety, depression, and stress after the course—results that were maintained a full year later.

Intrigued, I attended a weeklong mindful self-compassion retreat Kristin and Chris ran in the rustic and gorgeous mountains overlooking the coast in Big Sur, California. I found that compassion directed toward yourself adds a special, nurturing ingredient to self-awareness: *something to do* when painful or negative experiences arise. Mindful self-compassion combines the proven health benefits of mindfulness with something active to do with one's fears, concerns, and self-doubt. For me, that made mindfulness even more powerful.

While being compassionate with our children may be second nature, we don't always have the instincts to do the same for ourselves. How do we practice self-compassion? By simply sensing and acknowledging a challenge or difficult situation, recognizing that challenges are a shared human experience and that we are not alone, and then offering ourselves kindness and recognition, as we would to our children when they are struggling.

Kristin and Chris taught me a brief, research-backed exercise: the self-compassion break. It includes three acknowledgments that you say to yourself whenever you need or want to. I use the self-compassion break as a mantra that I repeat several times when I need to give myself support during difficult parenting (or otherwise difficult) moments. I hope you find it helpful and feel free to use your own words to get to the sentiment that resonates for you in each step.

Practice: The Self-Compassion Break

1. Notice and acknowledge a difficult moment, situation, or problem, and say to yourself, "This is hard," or, "This is stressful," or even, simply, "Ouch."

2. Remind yourself that you are not alone in suffering, and acknowledge or say to yourself, "I'm not alone," or, "This is how it feels when people struggle in this way," or, "All parents suffer at times."

3. Offer yourself kindness in some way, such as saying silently, "May I be kind or gentle to myself," or, "May I give myself what I need," or even asking, "What do I need in this moment?"

You can say these three statements to yourself any time things begin to feel shaky—or whenever you want. Consider also adding a light physical touch: briefly place your hand over your heart or touch your cheek, if that feels comfortable. These simple, quick gestures tell your nervous system that you are okay. They give you a message of safety, just as you would for your child. These three simple sentiments—*This is hard, I'm not alone, May I be gentle with myself*—can help us achieve calm during difficult situations. Supporting our own sense of compassion and well-being makes us better equipped to do the same for our children.

IS THIS NEW AGE OR SCIENTIFIC?

People sometimes react negatively to the idea of self-compassion. Some wonder whether it's a sign of weakness or a New Age form of self-pity. *To the contrary, we all feel judged as parents at some point, so practicing self-compassion is fierce, and brave, and sometimes counterintuitive.* Self-compassion isn't self-indulgent; it's a platform builder, and research shows that it can improve your physical and mental health. It improves your sense of well-being and makes it easier to repair the inevitable mismatches and ruptures we experience with our children. It's only human to mess up occasionally. Self-compassion also helps us model mental flexibility and self-acceptance for our children. Having said that, if applying self-compassion feels unpleasant or uncomfortable for you, in the spirit of this entire book—which respects every person's unique individuality—please be okay with that. Move on and try other activities that do feel nurturing to your body. There's no right way to soothe one's brain and body except for what works for you.

THE BREATH

Another powerfully beneficial parenting hack that doesn't require extra time is changing the quality and the awareness of our breath. When we control our breathing, we can better access our green pathway, which helps us recover from stress. Therapist Deb Dana reminds us that the "breath is a direct pathway to the autonomic nervous system." We can use our breathing to help us feel less distressed. Research shows that various kinds of *controlled breathing* reduce anxiety, stress, and depression and also support the immune system. And slow, controlled breathing sends a message to the nervous system to calm down, reducing anxiety.

Dr. Patricia Gerbarg and her husband, Dr. Richard Brown, are psychiatrists who conduct research and training on the health benefits of slow breathing and other mind/body treatments for stress-related medical conditions. They describe how slow, gentle breathing is calming, helping to reduce anxiety, insomnia, depression, stress, and the effects of trauma.

What's most important to remember is that for many of us who run on overdrive, carrying tension in our bodies, controlled breathing can help us calm our nervous system. We will feel better and be better able to co-regulate with our children. Remember this when you have your arms full of groceries, your child is screaming for dinner, or the dog just escaped out the front door. It's also handy when you're finally in bed, exhausted and trying to fall asleep, but your body and mind can't relax. Controlled breathing can, for many people, induce a feeling of relaxation. I recommend practicing this exercise or something like it when you are not in the throes of a busy parenting moment and have a little time to yourself.

Practice: Slow Breathing

Take a breath by inhaling slowly (through your nose if possible), counting to yourself from one to either four, five, or six (whichever feels most comfortable), and filling your lungs with air so that you feel your belly expand. Now exhale and again extend the out-breath to the count of four, five, or six. Repeat for several breaths in a row. If you feel more relaxed and alert, it's because you have just shifted your physiology into (or further into) the green pathway.

To achieve an additional sense of calm—at bedtime, for instance—try extending the exhale to be longer than the inhale. It's important to experiment with breathing and, of course, do what feels most comfortable in your body. If controlled breathing triggers any distress or anxiety, then simply and compassionately adjust to find the pacing that helps you feel calmer, more alert, and more grounded. Importantly, however, if controlled breathing doesn't work to help you feel more relaxed, don't beat yourself up about it.

The great appeal of controlled breathing is that it's so readily available. Even when you're trying to soothe a crying baby or calm a whining toddler, you can always breathe. And controlled breathing reminds us to be patient with ourselves and our children, providing a reminder that it's okay not to rush through our life and parenting. *Try adding a few controlled breaths to the self-compassion break and you have a supercharged self-care tool at the ready.*

COMPASSION FOR YOUR OWN HISTORY

Our reactions to life and to our children can be triggered by things that lie deep in our own history, by how our brains made sense of early stressful experiences and the bits of memory they left—many of which are outside of our conscious awareness. That includes how our parents and caregivers met our needs and how well their actions matched up with our individual requirements for comfort and security. *For example, if your parents had a difficult time tolerating your negative emotions or misinterpreted, ignored, or punished you when you had them, those early experiences can influence how you tolerate your own child's negative emotions.* You may feel more triggered when your child has those kinds of emotions or goes through the developmental stage you did when you had those negative experiences.

Sometimes, our own histories make it difficult to tolerate the stress from watching our children suffer, making us feel uncomfortable rather than sturdy when they experience negative emotions. That was a big challenge for me as a parent because I sorely wanted to spare my children from the suffering I experienced as a child. But remember, not experiencing tolerable amounts of "good" stress may actually hamper a child's potential for resilience. I worked on developing awareness of and compassion for my own triggers with a trusted therapist who helped me become more resilient as a mom. She co-regulated with me so brilliantly, and I will be forever grateful to her. We've all heard about college students who have had too much done for them their whole lives and crash once they're on their own and forced to manage their own lives. Again, awareness, and especially compassionate awareness of our own histories and the adaptations we made to survive, will spare our children from having to relive the experiences we did earlier in our lives.

One parent told me that his mother could tolerate only positive emotions and completely tuned out when he was angry or aggressive in any way. As a result, he became adept at pleasing people by being polite but kept negative emotions well hidden. When he became a parent, he became violently upset inside and didn't know what to do when his son was angry or fearful. He was quite relieved when I explained that a child's negative emotions are "equal opportunity"—that is, a reflection of the child's range of expected human reactions. It was healthy and expected for his son to experience anger. When he realized that he learned to please his parents as an adaptive survival strategy, he became more tolerant and accepting of his own feelings, and even encouraged his child to express his own negative emotions. *I helped him do this by increasing his ability to link basic body feelings to a wider number of emotion words.* We call this developing *emotional granularity.* (Think of the level of detail: fine-grained or coarse-grained.) At first, when he was distressed, he could only say he felt "bad" inside. After a few months, he was more detailed in his descriptions, describing emotions such as anxious, angry, or envious. In turn, he was able to help his son do the same.

If you feel triggered when you're with your child, you might want to stop and observe, without judgment. You may be triggered by a past memory or experience that you're not aware of. If you feel that unresolved issues from your past are causing struggles with your child, then it helps to be accepting and not beat yourself up emotionally. Regret, shame, and guilt will only make withdrawals from your body budget, so self-compassion is crucial here. Through no fault of their own, many of our parents didn't have the benefits of the positive-parenting information that we have today. Your well-intentioned parents may have given messages that made you feel badly about or question yourself—messages such as "You're overreacting," "Be polite and respond when an adult asks you a question," or "Don't make a scene!" Simply being aware of those messages helps us not re-create the same experiences for our children.

ACTIVITY: Reflect on and write down how your parents and caregivers saw and accepted your needs and emotions when you were young. Did your parents or the adults around you tolerate a narrow or wide range of your negatively charged emotions when you were a child and teenager? List your observations and then consider whether areas of past pain are showing up in your parenting life.

Sometimes we act instinctively as parents for reasons that are beyond our awareness. Subconscious triggers can cause us to overreact or underreact to our children's behaviors, depending on what's getting triggered. The tools in this chapter help to bring awareness and support your ability to see your child's needs from your child's standpoint rather than reacting from your own projected worries or trauma. It's useful to remember that our own challenging behaviors, like those of our children, are subconscious adaptations: Our nervous systems did what they needed to do to protect us and help us through the difficulties of our own childhood.

If you find that you have strong reactions to your child's behaviors that may stem from displaced energy in your past, treat yourself with compassion. Adverse past experiences, such as being mistreated or living in difficult situations in which you witnessed the mistreatment of others, especially if it was ongoing, can cause toxic stress and trauma. Help is always available to heal the wounds of the past. If you have challenges to your own regulation—especially if caregivers mistreated you or you experienced or witnessed violence—it's important to explore these matters and sort out how they affect your life as a parent. We can learn to observe and make sense of our automatic reactions. *If you feel that your past is burdening you as a parent, it's worth consulting a mental-health professional to help sort out the strong feelings evoked by parenting.*

If you ever find yourself feeling hopeless and battling thoughts of serious depression or self-harm, this is a serious medical condition that is treatable. The national 24/7 suicide hotline is staffed by caring professionals who can help you and provide hope. The U.S. number is 800-273-8255.

IT'S AWARENESS THAT COUNTS, NOT HAVING A PERFECT CHILDHOOD

Nobody has a perfect childhood. No matter our childhood circumstances, what makes a difference is *how we make sense of our life*—how we tell our story. Our children benefit when we have awareness of our emotions, physical sensations, thoughts, and memories. That awareness helps us to distinguish among our own stress, our personal issues, and our child's needs and emotions. It also prevents us from taking things out on our children.

We need not be haunted by our past. Healing comes from making the unknown known with gentleness toward yourself, so that past experiences no longer cause you to act unpredictably toward your child but rather signal that you are triggered and need to take care of yourself. We and the people who were charged with caring for us were trying to survive. We all are.

Gaining awareness of our own history and triggers can make the difference between our doing or saying something that frightens our child and something that connects with our child's underlying needs. We become aware of how we can anchor ourselves, realize when we're not regulated, and discover our own tools to get back.

PRACTICE: The next time you're in a difficult parenting moment, pause to consider what color pathway you are on that's reflected in the look on your face, the tone of your voice, and your gestures. *Ask yourself: "Is this moment triggering something in me?"* If so, don't judge yourself; rather, make a note to reflect on it when you feel comfortable doing so. Recognizing when we are triggered is the best way to prevent our past trauma from affecting our children.

WHAT BRINGS YOU INTO CONNECTION WITH YOURSELF?

Since stress is an inevitable part of our lives, it's important to have tools to help us calm our bodies. We all perceive stress differently, so it's useful to create your own supply of stress busters to help calm and restore your body budget. Each of us is unique and experiences the world differently, so we need to create our own toolboxes for managing stress and finding connection with ourselves. Determining what our toolbox should contain requires time for discovery. Have a journal or notebook handy to keep a list of activities or moments that feed your body and mind. Beware the negativity bias. I encourage parents to write down the activities that help them feel centered because, as one day gives way to another, the good moments can slip away while the stressful ones linger in our memory. Jotting them down in the moment helps capture them and then you can work to replicate them with intention.

ACTIVITY: Write down your personal strategies to feel more connected to yourself or to others. Perhaps it's getting the privacy of a hot shower without a child coming in and asking a question. Or maybe it's listening or dancing to music, taking a short walk during your lunch break, meditating, gardening, taking a yoga or spin class, or grabbing coffee with friends. Maybe it's allowing yourself simply to sit outdoors with your child, relishing the moment as the sun glistens off their hair. You and your nervous system know best what helps you to come back to balance. What are some positive experiences or habits that bring you peace, comfort, or joy? What activities fill you up when you are depleted?

Amelia Faces Her Past

Armed with these strategies for parental self-care, let's revisit the two families from the beginning of the chapter. Amelia, the mother who was so triggered by her son Silas's crying, and Tyrone and Dana, parents of Jaheem, whose struggles with each other and their son escalated during the COVID-19 lockdown.

To make sure that Silas's crying didn't have a physical basis, I contacted the referring pediatrician. When his pediatrician established that his crying spells didn't signal a medical cause for concern, Amelia and I began to explore the deeper roots of her reactions. I explained to her that sometimes an interaction, behavior, or developmental stage in our child can trigger our own stressful subconscious memories. All parents at some point find themselves triggered by their child's actions just as we all cycle through the blue and red pathways, depending on what life throws at us. What we do have control over, though, is how we think about these memories and what we do about them.

As we got to know each other and developed trust, Amelia explored her history as she tried to understand why Silas's crying was so challenging for her. When she'd had intense emotions as a child, she recalled, her mother, a supervisor with long hours at a grocery store, had done everything in her power to make Amelia happy as quickly as possible and to help her stop negative emotions. As a child, Amelia didn't have the space to recognize her own emotions because her mother didn't have a wide range of tolerance for *her own distress.* So little Amelia adapted, trying to help her mother by repressing her own negative emotions.

As a result, Amelia became adept at helping others feel safe and happy. That trait won her many friends and earned her an excellent reputation at work, but it came with costs to her own well-being. Amelia realized that her mother had had no outlets for her own emotions because she was so busy taking care of everyone else and making sure they had food on the table, shouldering so much pain through her own difficulties and challenges. She learned to mask her true feelings.

That helped explain why Amelia developed a cheerful, happy exterior but was easily triggered into distress herself.

Through our work together, Amelia came to understand that her mother had her own challenges with processing negative emotions—problems that remained in the adult Amelia's early and subconscious memories, triggering her safety-detection system when Silas cried and couldn't be soothed. I helped her create new meanings out of the sensation of hearing her baby cry. Just as a parent helps the child integrate experiences as tolerable and safe, I often find myself in the same position, helping parents interpret their own distressed feelings in new and compassionate ways. This helped Amelia curb her harsh self-judgment by demystifying her anxiety and shining a new light on it.

Eventually, with awareness and great self-compassion, Amelia grew more attuned to herself. She began to acknowledge the burdens and worries she carried. She constantly worried about being her family's sole financial source, fretted about taking time off for her health-care needs or Silas's, and stressed about the endless details she juggled daily at work and at home. She brought awareness to her triggers, feeling relieved that there was a reason for the seemingly irrational ways she interacted with her young son. She also had a deeper understanding of the complicated dynamics with her mother, who was Silas's biggest supporter. She worked on becoming more honest and assertive with her mother, and increasing her tolerance for Silas's negative emotions, sometimes phoning me for extra support. She also began to narrate her own experience to herself and used the self-compassion exercise, which allowed her to soothe herself and to know that she, as all parents, is not alone in her suffering.

After a few months, Amelia's tolerance for her own distress increased. She judged herself less. At one session, she told me she felt like she was pulling the shade up in her emotional window and letting the sun in. Not surprisingly, when she tuned in to her body and held herself in the same gentle regard that she held Silas, she stopped feeling so troubled by his crying.

Tyrone and Dana Find More Balance

When I met with Tyrone and Dana, both parents were experiencing states of depletion and in need of deposits to their body budgets. I reassured them that their son's behavior—his repeated interruptions during the workday, his inability to comply with their requests—was well within what we would expect from a child who loves his parents and is just starting to develop self-regulation. I also explained that his challenging behaviors reflected the stress the pandemic was taking on him, how he was missing his friends and teacher.

What they needed was some time to figure out how to make deposits back into their body budgets. Tyrone missed working out at his gym, which had to shut down because of the pandemic. So he used his year-end bonus to buy a stationary bike and started exercising again. He also began a weekly Zoom check-in with friends he had been missing. Dana realized that the main source of her stress was suddenly having to homeschool Jaheem. Fearing that he might fall behind in his education, she admitted that she had placed a great deal of pressure on him—and herself. As a result, she and Jaheem were riding the same emotional roller coaster. When he felt happy, so did she. When he was agitated, she began to feel agitated as well. Working from home had its benefits for Dana, but she needed to find ways to calm herself so that she would be able to help modulate Jaheem's shifting pathways rather than absorbing his emotional states. She found mindfulness helpful. After she started using an app with short meditations, she was better able to manage her stressful reactions. She also realized that routinely staying up late bingeing on TV shows was causing her chronic sleep deficit. Addressing her sleep requirements helped restore her tolerance and equip her to better attune to her son.

Finally, the couple realized that they hadn't had a date night in nearly a year. Trading babysitting with a family friend in their COVID pod, they started taking a weekly dinner picnic to a park just for the two of them to get some restorative time together.

These families found relief and answers the way all parents can: through compassionate awareness of our physical and emotional needs and our histories, and by allowing ourselves to seek comfort from those we trust. Now let's move on to parenting solutions that help us uncover insights about what's affecting the state of our nervous systems and how we can effectively co-regulate with our children.

RESILIENCE-BUILDING TIP: The most important tool in our parenting toolbox is our own emotional and physical well-being. But that doesn't mean we have to be perfect; the key is developing the awareness to identify your needs, finding self-care strategies that work for you, and having compassion for yourself as you do so. Valuing your own mental health and the ability to feel emotionally stable is one of the best things you can do for yourself and your child.

6

Making Sense of the Senses

HOW EMOTIONS ARISE FROM THE BODY'S EXPERIENCE OF THE WORLD

There will never be another you.

—Dr. Edith Eva Eger

Wouldn't it be great if your child came with an instructional guide telling you how to parent your bundle of joy? In fact, there is one: your child's body. As I've mentioned, sometimes our parenting efforts miss the mark because we're too focused on the child's thinking, their willpower, or the self-control that they simply haven't developed yet. No matter how many incentives or consequences we offer, or how much we try to reason with the child, things don't always turn out the way we'd hoped. If we want to understand what our child's behavior truly means, we need to pay attention to their *body-based* reactions to the world, which offer a wealth of information. One of the best ways to support our children is working to understand how they experience their world—and how that informs their basic feelings, behaviors, thoughts, and emotions. We start by understanding that the way we make sense of the world is through the sensations we experience.

Here's a simple example of why it's important to pay attention to how a child's body responds to their environment. Many children have extreme reactions to seemingly benign experiences, such as having their hair shampooed. Some children dread it, and the experience triggers red pathway behaviors such as crying, yelling, and pushing a parent away. Faced with such a situation, many parents (including me, back when my children were young) simply try to "power through," or suggest the child is making a big deal out of nothing. But both of these responses can further stimulate the child's threat-detection system. Powering through doesn't honor the strong signals coming out of the sensory experience, and using reasoning or logic at that moment doesn't usually help the child calm down.

I'm not suggesting you shouldn't wash your child's hair if they protest. What I am suggesting is that you do so in a way that reduces the child's stress and therefore the cost of their subjective experiences to their body budget. How to accomplish this—to parent in a way that honors your child's felt experience and supports their platform—is the focus of this chapter. *We'll explore how we can learn about our child's psychology by understanding how their bodies make sense of the world.*

Sometimes these moments of learning can take us by surprise. I still remember when my husband and I planned a trip to Disneyland for our creative and inquisitive daughter's fifth birthday: When we excitedly announced our big birthday gift, though, we didn't get the response we expected. Our daughter shook her head, saying she didn't want to go.

My husband and I exchanged looks, puzzled. What kid doesn't want to go to Disneyland? It took me years to understand, but I eventually discovered that my daughter wasn't being stubborn or ungrateful. She knew herself well, and her choice made complete sense based on her prior experiences at Disneyland. It was her parents who needed to learn.

Why do our children behave the way they do? Again, behaviors represent the tip of the iceberg. To understand what's below the surface, we need to explore how the child's body is interacting with the world.

GETTING FAMILIAR WITH HOW THE BODY INFLUENCES BEHAVIORS

The way that children—and all human beings—experience and make sense of the world around them is through their *sensory systems*. As we've discussed, our child's behaviors are a window into how they perceive the world. To help us interpret their behaviors, it's important to understand the impact of physical sensations.

As we learned in Chapter 3, our *central nervous system* is made up of the brain and the spinal cord. The *peripheral nervous system* connects the brain with the rest of the body through a superhighway of neuronal pathways. As the brain responds to the stream of information delivered from our body, the body acts on directives from the brain. This constant, bidirectional conversation takes place within all of us, all the time. The body sends information to the brain and our brain processes that information and sends signals back down to the body, causing our actions.

Think of what happens when we hear a passing police siren and instinctively cover our ears, or when a child gets a feeling inside their body, sits down on the floor, and says, "I have a tummy ache." When a child experiences discomfort, internal sensors in and near the child's digestive system send signals via the body-to-brain pathway, enabling the child to (sometimes) become aware of and act on those sensations, in this case, sitting down and telling a parent, "I have a tummy ache."

When the body sends signals to the brain, it directs us to respond in a way that keeps our body budget in balance—in scientific language, to maintain *homeostasis*. It's worth noting that we have *far* more fibers that go from the body to the brain, with some 80 percent of the fibers carrying signals *to* the brain and only 20 percent carrying signals back *from* the brain to the body.

Yet we often overlook that it's the information flowing to the brain from children's bodies that influences how they feel and what they do. Paying heed to these body-up signals helps us to customize our parenting to our child's unique physiology and allows us to better understand how they feel in body and in mind. In short, we use this information to begin to create a unique instructional guide to our child.

This holistic view of child development—and the shift in understanding behaviors and emotions as meaningful reflections of the individual differences in how a child perceives the world—is still unfamiliar to most educators, pediatricians, mental-health professionals, and parenting experts. *Our culture hasn't widely recognized the bidirectional influence between children's brains and bodies.* As a result, it's common for parents to receive conflicting advice about their child-rearing from sources as diverse as pediatricians, social media influencers, parenting books, teachers, and well-meaning grandparents.

That's why it's important to remember that the best guidance comes not from a book or a website but rather from *your child*: from *their* unique nervous system, which serves as a road map to parenting and is primarily influenced by how they constantly take in the world around them.

SENSORY EXPERIENCES DRIVE BEHAVIORS AND BUILD TRUST IN ONE'S BODY

When we understand that behaviors are the tip of the iceberg and have meaning in the brain and body beyond what we can readily see, how can we use this information to better support our children? One important way is to focus on how humans interpret and understand the world through our senses. When I was an undergraduate at the University of Southern California, little did I realize that on that very campus, a professor named A. Jean Ayres was developing a theory that would later become a pillar of my work as a child psychologist. An occupational therapist, psychologist, and researcher, Dr. Ayres studied the role of sensations in human behaviors and developed the field of *sensory integration*, which explains how the brain takes in and organizes sensory information, allowing us to respond adaptively to what's required in daily life and to engage with and respond to the world in a well-regulated way.

In the words of Dr. Ayres, "You can think of sensations as 'food for the brain'; they provide the knowledge needed to direct the body and mind." Signals travel from the brain and back down to our bodies, triggering corresponding behaviors. Dr. Porges, creator of the Polyvagal theory, explains it this way: "Sensory experiences drive our behavior and contribute to the organization of our thoughts and emotions." They do this because our past sensory experiences inform our reactions to similar experiences in the future. Lisa Feldman Barrett, the neuroscientist, tells us that our brains are incredible simulators, using our "past experiences to construct a hypothesis—the simulation—and [comparing] it to the cacophony arriving from your senses."

In this way, sensory processing underpins all human feelings, emotions, thoughts, and behaviors. As the way our children understand the world, it deserves a prominent place in parenting literature as well as in education and pediatrics. *Understanding how our brains use the data coming up from the body has had a profound impact on how I parent and practice psychology.* Of course, there are far more complex explanations about the sensory systems and the brain than I share here, and I urge you to seek them out if you are interested.

What I am presenting is a highly simplified version, along with stories from my work with families that I hope will help you understand how your child's sensory systems impact their behaviors and feelings, and how they can help inform our parenting decisions.

OVERREACT, UNDERREACT, OR CRAVE

First, it's essential to know that we don't all process sensory information the same way. As we've discussed, one child's safety system might interpret a particular sound as threatening, while another's system deems it safe and enjoyable. Some children are *over-reactive*, experiencing a sensation more powerfully than most, while others are *under-reactive*, registering an experience less powerfully than most. Some have *sensory craving*—they can't seem to get enough of a particular sensation, so they seek it out repeatedly. Still others can tolerate a wide range of experiences. Variability is to be expected in our sensory experiences of the world.

Does your child fall into one of these categories? You might try using the "so" approach in Chapter 5 to help assess your child's sensory processing. Over the course of a week or two, *stop* and *observe* your child's reactions to various sensory experiences.

Though we learn about the five senses in school, there are actually eight sensory systems that contribute to how a person takes in the world, informing our behaviors, emotions, memories, and relationships. As I introduce each system in the following pages, I'll describe ways to reflect on how your child's senses are informing their behaviors and how to use each sensory system as a guide for reestablishing calm in your child or yourself. *(It's useful to keep a journal and jot down your observations and ideas.)*

Of course, our bodies are always involved in a rich *multisensory* experience. We are exposed to sounds, smells, sights, tastes, and all other sensory stimuli *simultaneously*, and our brains are constantly taking in bits of information, comparing it to past experiences, piecing it together, and *integrating it—all* at the same time. In other words, all of the sensory systems influence one another. Think of biting into a juicy red apple. You are tasting, seeing, smelling, looking, touching, and moving in synchrony as you enjoy eating that apple. That's why Dr. Ayres calls it *sensory integration*: It's a highly complex and integrated system that helps us understand the world and maintain stability in the body.

I learned about multisensory processing from my mentor, Dr. Serena Wieder, whose work with Dr. Stanley Greenspan beginning in the 1970s linked children's emotions and behaviors to their complex sensory experiences. Understanding the body-up contributions to children's behaviors was the missing piece that hadn't been addressed in my psychology training, which focused on analyzing and altering people's behaviors and thoughts. When I was introduced to how multisensory processing contributes to child development, I stepped into a new way of supporting children and families.

For example, it's common for parents to come to me because their three-year-old child's preschool has deemed their challenging behaviors too aggressive and kicked them out. Often, I discover that high reactivity in multiple sensory systems has propelled the red-pathway fight-or-flight behaviors of such toddlers. Provided with sufficient co-regulation from caring adults who support them with this in mind, these children eventually build new meanings of their sensations, enabling them to cope better.

Our sensory systems operate simultaneously and influence one another, but for the sake of simplicity I will provide a basic description of each system separately and how it can influence a child's behaviors within this great symphony.

Let's start with a sensory system that I mentioned in earlier chapters, one with which you likely aren't familiar: *interoception*.

It's the most important one because it helps us understand how the world *within* our bodies gives rise to our most basic feelings, impacting our emotions and behaviors.

INTERNAL SENSATIONS (INTEROCEPTION)

Internal sensors send information to the brain about how our bodies feel on the *inside*. *Interoception* refers to sensations that provide information about the internal state of our body. Interoceptors located near and in our internal organs automatically send important information to the brain to help our body stay in balance and regulate our body budget. When we *notice* feelings from our insides, that *interoceptive awareness* triggers conscious reactions such as hunger, thirst, twinges, pain, or feeling "butterflies" in the stomach.

Interoception also affects our basic feelings and moods, and that's why we need to know about it as parents. Researchers and therapists have linked interoceptive awareness to emotions, and emotional regulation. Indeed, they are suggesting a groundbreaking idea: that interoceptive sensations *generate* our most basic feeling states and moods and, sometimes, what we come to label or identify as emotions. This makes so much sense to me as a psychologist. Even before I knew the term *interoception*, I found that the *awareness of sensations* was key to helping children understand their emotions and behaviors. *In my clinical work, I have found that children with better awareness of bodily sensations also have better self-regulation.* We can learn a great deal by how a child takes in the world through their senses and the benefits and costs of those experiences. In my experience, *helping children (and adults) observe and make sense of their bodily sensations is one of the best ways to support self-regulation.* Eventually, we can help children label those sensations with emotional or other descriptive words.

Stop and Observe:

- *How comfortable is your child with identifying and naming basic sensations coming from within their body? Of course, we would expect this from children once they have developed the ability to answer such questions. Does your child seem to notice if they are hungry or have a growling tummy? Can your child tell if they're sleepy and want to go to bed or take a nap? Can your child tell you if they are in pain, where the pain is located, and the level of the pain?*
- *Does your child have a pattern of negative reactions to internal sensations? Do you notice distress, high reactivity, or behavioral challenges when your child may be constipated, thirsty, or hungry, or has other uncomfortable sensations emanating from inside the body?*
- *When your child notices sensations, does it help calm the child and enhance relational connection? If a child can notice sensations, especially unpleasant ones, and talk about them or come to you for support, they're on the way to gaining self-regulation. Want to help your child to do this? You can model it for your child by naming your* own *bodily sensations and how you experience them, as they happen in everyday life: "I'm hungry, I'm going to have a snack," or "I'm thirsty, I want a nice big cup of water." We should also let children know that whatever sensations they feel, it's okay; these are clues to what's happening in their bodies and what they can do to feel better.*

Interoception Case Study: Time to Go

I once had a young patient who almost always felt the need to visit the restroom twice or even more frequently during our forty-five-minute sessions. She simply felt the urge strongly and often. Around the same time, I had another patient who experienced the opposite issue: He would wait to go until it was too late, so his mother had to monitor how long it had been since he had gone to the bathroom and remind him to go.

Why did one child visit the bathroom frequently, while the other neglected to go at all? The way each child experienced the same sensation inside their bodies—the urge to urinate—was different. One child experienced a low level of sensations coming from his body. His interoceptive awareness was *under-reactive*, so we worked on helping him to tune in to his body sensations and pay more attention to its signals. He eventually developed more awareness and was able to steer clear of accidents.

We helped my client with strong or *over-reactive* interoceptive awareness to develop a greater tolerance for what she experienced as the urgent signals that compelled her to rush to the restroom, even though she had gone ten minutes prior. Her parents and I, under the guidance of her pediatrician and an occupational therapist, used play to explore and model sensations, like sleepiness, hunger, thirst, or the need to go to the bathroom, in our pretend-play characters. We read books, wrote stories, and drew pictures of what sensations felt like in our bodies.

For both families, I encouraged the parents to help their children to explore and talk about body sensations. This approach not only helped them resolve their issues with pottying, but I assured them that along the line, this would also help them feel more comfortable talking about and labeling basic feelings and emotions as well—a great bonus!

Interoception Case Study: From Vague Pain to Compassionate Awareness of Inside Sensations

Another client was twelve-year-old Kira, whose pediatrician sent her to me for symptoms of worry and anxiety. Her parents told me that she had been plagued with stomachaches since she was a toddler. Two different gastrointestinal specialists had found nothing medically amiss.

Kira is quiet, shy, and well behaved—the kind of child who is often misunderstood as being "fine" because of her calm demeanor. But that didn't tell the whole story. Inside she was anything but calm. She had strong interoceptive awareness from her gut and told her parents how frustrating it was to keep going to doctors and being told that they couldn't find anything wrong. My work with Kira centered around helping her tune in to her body sensations in a new way.

Validating that every feeling from the inside is valuable information, I helped her develop a larger vocabulary than "It hurts" or "It doesn't hurt." She held these sensations in a sensitive body that overreacted to many other sensations, including sounds, full sun, and bright lights. What helped her was enabling her to become friendlier toward the many sensations her interoceptive awareness was bringing up. We worked on welcoming her body's sensations with less distress and becoming curious rather than fearful about what the signals meant—whether it was the need to eat or drink, to move her body in certain way, or to seek comfort and connection from her parents. Remember the concept of emotional granularity, and the dad who had difficulty tolerating his son's negative emotions? Kira, too, had a hard time linking words to basic feelings inside her body. As we worked together, she developed a larger vocabulary for what she was experiencing. She began to tune in and talk about her interoceptive feelings with more *emotionally descriptive words*, based on how she felt in the moment: calm, jittery, happy, playful, scared, nervous, excited, etc. Kira needed support to better understand and sort out those strong signals that came from deep inside, and over the course of several months she experienced stomach pain less often.

Now we'll examine the other sensory systems that send feedback to the brain from the more commonly known senses, and stories about how sensations contributed to concerns about children's behaviors.

HEARING AND THE AUDITORY SYSTEM

We are constantly bombarded by sounds, in both the foreground (a person speaking close by) and the background (traffic sounds, music emanating from speakers in a busy supermarket). As sensors in the inner ear send information to the brain, they integrate with other forms of sensory information, enabling us to make sense of the various sounds we hear.

We each experience sounds differently. Often, adults are able to control aspects of the sounds we hear—the volume of music playing in the car, or even the type of music we prefer to listen to. Since children generally have less opportunity to exert that kind of control, their reactions can show up in their behaviors.

Children might not even realize they are over-reactive to sounds, but as parents, we can observe behaviors associated with such difficulty: a noisy restaurant might trigger a meltdown, the hum of an air-conditioning unit might cause a child to become distracted. Conversely, a child who is under-reactive to sounds might seem not to pay attention, ignoring instructions at home or at school, unless the source of the sound is at a certain volume and directly in front of him.

Stop and Observe:

- *Notice your child's patterns of reaction to various sounds. How does your child react to everyday sounds such as the washing machine, a fan, paper crumpling or tearing, a vacuum, or sirens? Observe variations in the volume and tone of different sounds.*

How does your child behave in places with background and foreground sounds, such as a shopping mall or gymnasium? Does your child prefer some kinds of music over others?

- *Do you notice a pattern of negative or challenging behaviors when your child experiences specific sounds? Do some sounds seem to make your child fidgety or even upset regularly?*

  If similar sound experiences seem to precede a consistent pattern of red- or blue-pathway behaviors, that might indicate that certain sounds trigger their safety system and you are observing the child's protective response.

- *What kinds of sounds calm your child or bring happy, joyful interactions into your relationship? The soft, melodic tone of voice that we use intuitively with babies is naturally calming for many. Try using it and see if your child settles down. Even if the child isn't a baby, you can still use an adult voice laden with cues of safety by varying your tone. Even better: Compassionately mind your own platform to see if you are feeling calm enough to co-regulate and witness your child's distress without judgment.*

Notice your child's emotional reactions to your various tones of voice. Tracking your observations in a journal can help you gauge the vocal tones and qualities that most help your child feel safe, enhancing co-regulation and enabling connection. Sounds can be calming or distressing to a child. Qualities of the voice help us to judge whether it's okay to draw closer physically and emotionally. That's why children pick up on our *emotional tone* before they register our words. They're naturally sensitive to the emotional aspects of voices, so it's important to pay attention to how your tone of voice is landing in your child.

Just hearing a beloved parent's voice can make a child feel safe. When my youngest daughter turned three, her older sister moved into her own bedroom. Having shared a bedroom from birth, our youngest was both excited and hesitant. To comfort her, at bedtime, we would all say good night through our open bedroom doors. The sounds of our voices made us all feel safer and more connected, helping to calm all of us.

Then there are the sounds we don't even think about often: the hum of the air conditioner or heater; the mechanical sounds of a mall's elevators and escalators; restaurant background music—all sending messages to the nervous system and in turn triggering behaviors. Even when these sounds go unnoticed by parents, they can destabilize the platform of a child with particular sensitivities to sound.

Hearing/Auditory Case Study: The Gruff Grandpa

One couple told me that their child's grandfather felt hurt that his eight-month-old grandson repeatedly rejected him. Whenever the grandpa paid a visit, the baby cried and fussed when the grandpa held him or spoke. The poor man couldn't help but take it personally.

I spoke with the family, and they shared a short video of one of the grandfather's visits. I suggested that the parents track the infant's reactions to different types of sounds. Reviewing it, I discovered that the baby was sensitive to lower-frequency sounds. So I suggested that the grandfather try altering his voice just a bit, perhaps adding a more singsongy quality and speaking more quietly around the baby. Within days after he made the change, the baby stopped crying when he visited. The baby's system was now predicting and interpreting the grandpa's voice as safe instead of triggering. And in time, he remained comfortable around his grandfather, even when he went back to using his more natural voice.

SIGHT AND THE VISUAL SYSTEM

Two different types of sensors, cones and rods, send information from the retina of the eye to the brain's visual-processing centers. But each of us processes visual cues in our own unique way. For example, how do you and your child react to sunlight? What about fluorescent lights? Does it bother you when things aren't in the right place? Do you tend to adjust picture frames on the wall if they seem off by a few centimeters? If so, your nervous system probably enjoys the predictability of that visual organization.

Similarly, some children experience distress if things in their environment are rearranged.

A nine-month-old I knew could tell when something was added or missing from her bedroom. She automatically surveyed a room, staring at anything new that appeared, as if trying to figure it out. She would then crawl to the new object and touch it. Her visual system was very active. Her older brother, on the other hand, seemed oblivious to the same kinds of changes. They simply had different ways of understanding the world visually—and in combination with the other senses.

Stop and Observe:

- *Notice your child's patterns of reactions to various sights. How does your child react to seeing your cheerful or stressed face? Does your child prefer bright lights or soft lighting? Or perhaps your child ignores such details. Does your child prefer to look at moving or stationary objects? Do you notice any patterns of reactions to what your child sees in the world—negative or challenging behaviors that routinely occur when your child sees certain things? What are the specific visual triggers you observe?*

- *What kinds of sights* calm *your child and bring joyful interactions into your relationship? Are there certain picture books that your child loves to read with you? Sitting in your lap, the tone of your voice along with the drawings or pictures are likely a winning combination for your child's overall sensory experience. What we see can also bring comfort and stabilize our body. One child I worked with brought a laminated photo of her parents with her to kindergarten to look at when she missed them. It served as a visual reminder of her beloved family, a visual aid to regulate her emotions.*

  Notice how your child responds to you when you're tense in body or face. We often "wear" our platform on our visage, transmitting our emotions through our facial expressions. That's another reason to check yourself and give yourself what you need so that you can soothe your child through your empathic facial expressions. You and your child will both benefit.

Sight/Visual Case Study: No Green for Me!

I worked with a six-year-old child, Gerard, who didn't like to eat anything green. When he was two, he gagged at his first few tastes of broccoli, then refused to eat peas, green beans, or anything of that color. His mother worried that he might avoid green vegetables forever. This posed a particular challenge since the family had a vegetarian diet!

I understood Gerard's parents' concern about his nutrition, but their approach hadn't proven effective. In short, they'd told him, "Look, vegetables are good for you, so you need to learn how to eat them—get over it." I gently reminded them that the safety system thrives on compassion, so they shifted to another strategy: finding ways to help their son connect positive feelings with green foods. I also explained that look, smell, and taste can trigger immediate reactions and limit a child's willingness to try whole categories of foods.

Instead of focusing on convincing him to eat, we devised an enjoyable way for child and parents to explore food together, starting with touching and smelling a range of different foods, and then adding in green foods, and then having Gerard pretend to feed green food to his superhero action figures. Gradually, and over time, Gerard moved from avoiding green food to associating it with safe, connected feelings—thanks to his parents' choice to help him in powerful yet indirect and nonthreatening ways. As he formed new, positive memories, Gerard was willing to integrate healthy green (and other colored) foods into his diet.

TOUCH AND THE TACTILE SYSTEM

The tactile system is our largest sensory system, covering the entire body, sending information from sensory receptors to the brain. Your own tactile preferences are evident in your preferences for clothing, fabrics, bed linens, and towels. Some people tolerate and enjoy (or don't notice) a wide range of textures, while others have over-reactivity or under-reactivity to certain kinds of touch.

Stop and Observe:

- *Notice how your child reacts to different types of touch or sensations on the skin. Does your child like having their hair washed or brushed? How does your child react to touching different textures, such as clay, dirt, or soft foods? Do they enjoy play activities such as finger painting, or touching squishy or hard surfaces? Do they prefer to wear the same clothes repeatedly? Perhaps it's because those clothes feel better, don't have annoying tags, or are of a certain texture of fabric. Certain kinds of touch experiences can make some children feel uncomfortable or even anxious.*

- *Do you notice a pattern of negative or challenging behaviors when your child touches certain substances, fabrics, or foods? Write down the* textures or objects *to which your child has a strong negative reaction.*
- *What kinds of touch calm your child and bring joy or joyful interactions into your relationship? Do they prefer strong hugs or light embraces? Do they enjoy a massage on the back, shoulder, or arms? Many toddlers and young children have blankets or soft pieces of fabric or stuffed animals that they hold or touch in order to soothe themselves. These children are regulating themselves via the sense of touch; they get comfort from the sensations coming from what touches their skin. The next time your child is upset, consider offering a soothing touch, customized to the child's preferences.*

Touch/Tactile Sense Case Study: The Rejected Mom

I worked with a toddler who quickly pushed away his mother if she unexpectedly stroked the child's arm or face. The toddler seemed to avoid physical affection from his mother, who sometimes felt hurt and confused.

I assured her that her son wasn't rejecting *her* but adapting to *his* tactile (touch) over-reactivity. Some kinds of touch simply felt uncomfortable on his skin and he was reacting to his sensory preferences. When it comes to touch and textures, I explained, we all have unique sets of preferences. She took the opportunity to experiment with various kinds of touch (with his permission), including stronger hugs, embracing the child when his body was facing away from her rather than toward her, and shoulder rubs instead of a light touch on the skin. By accepting his preferences and working to discover his likes and dislikes, her connection grew stronger and she stopped feeling rejected by his tactile preferences.

TASTE AND THE GUSTATORY SYSTEM

Our taste receptors, mostly located on the tongue, help the brain experience the sensation of taste. Think of your favorite foods. What comes to mind? Our sense of taste, like all the other senses, is personal and imbued with emotional memories from past sensory experiences. Foods begin to take on personal meanings over time. We associate some foods (like my mother's rice pudding) with comfort and pleasure, and others (such as foods that made you sick in the past) with negative reactions. Taste is also strongly integrated with our olfactory (smell) system.

Stop and Observe:

- *What do you notice about your child's food preferences? Do they prefer salty, sweet, spicy, or bland foods? What foods does your child love and request again and again? The sense of taste is associated with the other senses, such as sight and smell, so you might observe your child having big reactions to simply seeing or smelling a certain thing.*
- *Does your child have a pattern of negative reactions to certain foods? Is mealtime a constant struggle? List the foods your child struggles with (if any). What about food texture—does your child*

gag or have difficulty swallowing foods with certain textures, such as smooth pudding or crunchy chips?

- *What kinds of food experiences does your child enjoy—and do you have joyful interactions with the child around food and meals? Are you able to slow down and enjoy mealtimes with your child at least once a day? Although feeding children can sometimes feel like a task, meals are also ideal opportunities for communication and fun. If a child has a negative reaction to a new food, openly acknowledging that is a good start. Being calm and encouraging—even playful—can help to co-regulate a child to the green pathway where they are much more likely to try foods with new flavors and textures.*

Taste/Gustatory Case Study: The Daddy Café

A colleague told me that his two children had such wildly different preferences in food that he often felt like a short-order cook. His daughter loved spicy foods, while his son preferred bland dishes—and foods that were soft, not crunchy. By accommodating his son, the dad wondered whether he was spoiling the children and even encouraging him to be a picky eater.

I appreciated how observant the father was and, after having a meal with the family at their home, I assured him that he was witnessing a biological preference having to do with taste—which wasn't necessarily the same as pickiness or stubbornness. I assured the father that with some patience and co-regulation, he could gradually help his son to expand his food choices. The dad slowed down the process and gave his more sensitive eater more opportunities to try new things when the son felt like it, providing him with the support and validation that helped increase his tolerance for trying new foods.

Once the father grasped that his son wasn't choosing to be a picky eater, but rather had body-based reactions to the smell, taste, and texture of spicy or crunchy foods, he changed his approach. He set out a few food choices instead of one and encouraged his son to try those that appealed to him. Instead of saying, "Please try it—I worked so hard to make this for you!," he gently observed his son's hesitancy with more compassion: "Hey, buddy, this is a new recipe. If you want to try, let me know, and if not, that's okay, too." *The solution wasn't necessarily making different foods for his child, but rather increasing the co-regulation around the child's reactions, helping his child create a new relationship with certain foods.* Interestingly, when his father became more patient, his son tried more foods. While his son never grew to love spicy food, he did begin to eat a greater variety of dishes, and eventually the dad stopped making separate meals for his kids.

SMELL AND THE OLFACTORY SYSTEM

Chemical receptors in the nasal structure send important messages that our brain registers as smells. Our sense of smell detects warnings of whether something is safe to eat. What's the first thing you do when you grab an item that's been in the refrigerator past its expiration date but that you want to eat? You smell it! And smells, like all the senses, are closely linked to memories. On my evening walks, I sometimes catch a whiff of onions frying through a neighbor's open window and I'm transported back to my childhood, when my grandmother would make a stew on weekends.

Stop and Observe:

- *How does your child react to different types of smells or fragrances? Do they notice or complain about smells? Do they routinely gag after smelling something? If so, these may be signs of over-reactivity to smells. Or perhaps they don't take note as readily as you or others do? That might mean they are under-reactive to smells. Children, like all of us, can have immediate positive or negative reactions to certain smells such as soaps, shampoos, foods, or air fresheners.*

- *Does your child have a pattern of negative reactions to certain smells? Does the child have strong reactions, such as refusing to eat certain foods or not wanting to visit places that smell a certain way—restaurants or the perfume section of a department store? Does your child have physical or emotional reactions to smells, such as gagging or becoming upset?*
- *What kinds of smells calm your child or bring enjoyment or pleasure? Does your child enjoy certain smells over others? Does your child seek out any particular smells? The next time your child has a reaction to a smell, try noticing it along with the child. Ask the child to talk about or try to describe the qualities of the smell (pleasant, unpleasant, good, or bad) and tell you more about it. It's a useful opportunity for the child to start developing awareness of sensations—a step toward noticing their emotions.*

Smell/Olfactory Case Study: The Dairy Farm Dilemma

A family I knew drove to California's high Sierras every summer. On the way, when they'd drive past a dairy farm with a distinct and strong manure odor, one of the three brothers would always gag. As a result, his brothers teased him mercilessly for being so sensitive to smells.

This child simply had an *over-reactive* sense of smell, while his siblings didn't. His smell system triggered an instantaneous alarm and gagging reflex (along with immediate unpleasant interoceptive awareness) and theirs didn't. My suggestion to the parents was to make an effort to teach the kids about sensory reactivity, making it clear that one's level of sensitivity isn't a simple choice. *This is a valuable lesson for all of us, because it's easy to make judgments about people's sensory preferences as character or personality flaws, when in reality it's those preferences that influence our feelings and behaviors.*

BODY AWARENESS AND THE PROPRIOCEPTIVE SYSTEM

Most of us are less familiar with the proprioceptive system. It's a remarkable system that tells the brain about our body positions. According to Dr. Ayres, it does so by sending information up to "the brain about when and how the muscles are contracting or stretching, and when and how the joints are bending, extending, or being pulled or compressed." In other words, it tells us where our various body parts are: standing or sitting, bending or reaching. Think about the last time you bumped into the side of a table edge or other hard surface because you guessed wrong by about an inch about where your body was in space.

Our muscles and joints constantly send information to the brain about our body position.

Proprioception is the sense that helps us to move efficiently and feel what we are doing. It's what allows us to button up a shirt efficiently. The brain receives feedback from the muscles and joints in our hands, enabling us to *sense* how to close the buttons without necessarily needing to *look*. Children with difficulties in proprioception sometimes have poor penmanship because they feel only weak feedback regarding how much pressure to place on the pencil or pen. Other times, children with an under-reactive sense of proprioception might get too close or push too hard when they are trying to play or be in a space with their peers. It's important to know where your body is in relationship to others and to things like furniture, yet we don't think about it much unless there's a problem.

During my first year of college, I contracted a serious virus that caused inflammation in my inner ear. It temporarily took away both my spatial awareness and my sense of balance. It was so strange not knowing whether I was standing up or sitting down. I experienced extreme nausea and had to close my eyes because it felt like the world was spinning. I couldn't eat, and I literally needed people to carry me because I wasn't able to discern my body position or even feel gravity. The only treatment available was for me to be heavily sedated in the hospital until the inflammation subsided after about a week. Fortunately, I recovered and experienced only minor long-term impacts, but it was the most disorienting experience of my life and has given me great compassion for the children I see with spatial awareness or balance challenges.

Stop and Observe:

- *Does your child move with efficiency, appropriately for their age and unique development? (Of course, babies and toddlers are still developing their abilities to move efficiently.) If your older child has to look at their hands or body in order to accomplish everyday tasks, such as tying their shoes or fastening buttons, then a feedback loop from the proprioceptive system may be still developing.*

- *Does your child have a pattern of negative reactions to situations that require body awareness, such as team sports or other physical activities that require feedback from the muscles and joints? Does your child seem to bump into other people or furniture, or misjudge how strongly to push or pull objects? Does your child feel the need to hang all over you? Some children instinctually drape their bodies on other people in order to feel more grounded and safe. Doing so provides more feedback to their muscles or joints, helping them to feel calmer. On the other hand, it could also be that your child just loves the feeling of that physical closeness! Sometimes children whose proprioceptive systems are still integrating with the other systems have an intensity to their interactions that can cause social difficulties. They might hug a sibling or peer too often or too hard. They might tap another person too hard when they intended just a light tap. These kinds of unintended actions can have a negative impact on a child's relationships with peers, who understandably misinterpret the intensity of the gestures as too strong or aggressive rather than as a bid for connection.*

If your child seems to have difficulty with body awareness, it's useful to offer more experiences that can help integrate that system with the others. Going to a fun play gym, using blankets to playfully roll up like "burritos," and playing games in which children "grade" various degrees of touch as light or heavy can help your child integrate their sense of proprioception.

- *What kinds of activities bring relational pleasure or help your child feel where their body is in space? Does your child enjoy climbing on jungle gyms, and can they interact with peers or you at the same time? Or perhaps your child needs help and emotional support to climb on structures or ride a bicycle—activities that require solid proprioceptive feedback. Does your child like to be wrapped in a blanket or play in one? I often encourage parents to experiment to see what kinds of reassuring pressure their child enjoys. You can pretend to make a human "sandwich," for example, if your child loves to be playfully wedged between two pillows or couch cushions. The more fun the child (and you) has, the better!*

Proprioceptive Sense Case Study: The Soccer Player's Son

I met the parents of a little boy who began having behavioral meltdowns after the first time he played soccer. His mother had played college soccer and had signed on to be his team's coach, feeling excited to share the game she so loved with her child. But she worried that her son might feel pressure to perform and wondered if that might have caused his behaviors or seemingly minimal effort. She also noticed that when the boy was running, he continually looked down at his feet, seemingly unable to keep his eye on the ball and falling down frequently.

One of the joys of parenthood is experiencing activities that you love with your child. But the process doesn't always go as expected. Exploring why the child needed to look at his feet as he learned soccer, the mother consulted with a pediatric occupational therapist, who found that the child had *under-reactivity* to his sense of body awareness in his trunk and legs. The child's proprioceptive system wasn't smoothly *integrating* with his other sensory systems, so he compensated by using his *vision* to know where he was in relation to the ball. In a fast-moving game like soccer, that wasn't ideal, and he fell frequently. That made him feel embarrassed and agitated, triggering his meltdowns. Far from enjoying the game, he found it stressful. During weekly occupational therapy sessions, the therapist taught his mom how to support the boy's growing sense of body awareness. Eventually, the child grew to love soccer, which became a perfect way to integrate his sense of body awareness with his other senses. With his mom continuing to coach his team and support him emotionally, the game became a satisfying bonding experience for the whole family.

BALANCE AND THE VESTIBULAR SYSTEM

We come to understand ourselves through our bodies, and the vestibular system, along with the others, helps contribute to our sense of our self, to know what it's like to live in *this* body. The vestibular system literally keeps us grounded. Sensors in the inner ear send the brain information about the position of the head and entire body in relation to gravity and movement, affecting our balance. This critical system lets you know where your body is in space, whether you are moving or still, how fast you're going, and where you are in relation to gravity. If you have ever experienced motion sickness, it was because of your vestibular system registering that information and making you feel unwell or dizzy. It's easy to underestimate how important this system is to feeling regulated. For example, many toddlers have difficulty tolerating long drives in car seats. When a child sits in a car seat, so many different senses are involved: the tactile and proprioceptive senses from the snug straps across the chest, the sounds of the car, the visual shifts of looking inside and outside, and all of the movement that provides strong input to the child's vestibular system.

Children can get a great deal of vestibular input (as well as visual and proprioceptive input) when they jump on a trampoline. (I prefer trampolines with side rails to protect against falling off.) Jumping is a great activity to challenge the vestibular system, while landing on a forgiving surface provides pressure on the muscles and joints, which is the closely linked proprioceptive system.

Stop and Observe:

- *Does your child have a pattern of reactions to certain movement experiences? Does your child crave—or avoid—certain types of movement, such as swinging? Some children avoid climbing or jumping, or playground equipment like swings and slides. Others get upset on escalators (or avoid them entirely) or suffer frequent car sickness. Sometimes, children become distressed when their head is tilted back as you rinse shampoo from their hair. These children may be experiencing challenges to the vestibular system.*
- *What kinds of movement calm your child and bring relational enjoyment? Does your child seek to move around? Does your child like to dance, be swung, or otherwise move their body during playtime with you? What are your child's favorite kinds of movement experiences? Or perhaps your child prefers to be more stationary and* not *move much when you are interacting and playing. When you know how your child's body experiences movement, you can consider that information as you find ways to connect, co-regulate, and have fun.*

Balance/Vestibular Case Study: Playground Struggles

A mother told me that when her five-year-old son was on the playground with other children, she often couldn't get him off the tire swing or the bucket swings. He never seemed to want to stop swinging, even when other children were waiting and his turn was up.

The child loved the feeling of his body in the air and on the swings. The problem: He couldn't get enough, so his mother and teachers had to manage his meltdowns when it was another child's turn. He understood the social rules of sharing, but his body felt *compelled* to swing. I observed that the boy experienced sensory *craving* in the vestibular system, and suggested that he engage in occupational therapy that included targeted play activities with his mom that would help to better integrate his vestibular system with his other senses—including his visual and proprioceptive senses. The therapist suggested some fun activities, including pulling a wagon of logs when he was in the yard, having his parents let him push their shopping cart when they were shopping, and playing with him on various kinds of equipment. Some children with sensory craving benefit from occupational therapy because it takes some trial and error—and titration that occupational therapists are trained in—to help them integrate vestibular sensations with their other sensory systems. After several months of occupational therapy, the boy didn't refuse when he was asked to relinquish the swing and enjoyed playing on a wider range of playground equipment and movement experiences with other children. *He felt calmer in his body, and what originally appeared to be a behavior problem (not sharing the swing) turned out to be an underlying sensory integration challenge.*

MODELING RESPECT FOR THE BODY'S SIGNALS

We can give children a head start in befriending their feelings and emotions by helping them identify the full range of all their sensations without judging them as good or bad. The body provides us with a great road map for parenting, but we don't generally call attention to it. Sometimes, with the best of intentions, we hint to our children that they *shouldn't* heed their bodies' signals. "You can't be that hungry—you just had a snack," we might say, or "It's just a small scratch, it shouldn't hurt too much." In situations like these we may think we are being supportive, but in suggesting they ignore their body's feelings we actually devalue the very signals a child should pay attention to in order to self-regulate.

When we try to convince a child that everything is okay when it's really not, we dismiss their reality—and the important signals their body is sending. This is especially true when the child's safety-detection system is somehow registering threat even though the child is objectively safe—such as one who goes ballistic when getting their hair shampooed. We don't want to override the child's distress but rather co-regulate emotionally and respect their body's reaction, and in doing so model respect for the *very foundation of our basic feelings, our bodily sensations.*

Most children process information through their senses without obvious difficulty. But when behavioral or other challenges start in early childhood and can't be easily explained, we should ask whether sensory-processing differences are a factor. If your child has extreme sensitivities, overreacts or underreacts to certain sensory stimuli, avoids certain sensory experiences, or routinely craves certain sensory experiences and experiences dysregulation when they can't have access to them, it's worth consulting a professional. An occupational or developmental therapist with training in sensory processing can evaluate your child and determine whether intervention might help.

GOLDEN OPPORTUNITIES TO BUILD RESILIENCE THROUGH AWARENESS OF BODY SIGNALS

Neuroscientists have recently found a strong connection between our ability to detect our body's sensations and our ability to regulate emotions. To a brain-body therapist like me, this is amazing news, though it comes as no surprise. For years, I've found that encouraging children to tune in to their body's signals and sensations helps them build self-regulation. *We help children develop emotional coping abilities by using gentle, nonjudgmental observation.* When we appreciate a child's sensory experience of the world, instead of using judgmental statements such as "They're too sensitive," we can be curious about the child's preferences and respect their sensory experiences.

Too often, we overlook the ways children perceive the world—and other people—through their senses. For example, we may feel embarrassed if our child ignores acquaintances or family members—or reacts to them in ways that seem inappropriate or abrupt. Instead of immediately telling the child to be polite or "act nice," we can co-regulate in a manner that's appropriate to the situation, and in doing so, demonstrate respect for the child's natural response. When we fail to do this, we can unintentionally cause the child—eager to please us—to mask their true emotions. On the other hand, validating the reaction offers an opportunity to build the child's resilience, promote self-advocacy, and encourage the child to trust their instincts. This is particularly important for girls, who are socialized early to accommodate the needs of others.

We can then understand that most children don't *choose* to feel things deeply or strongly or to be picky; it's much more complicated than that. When we help a child learn to tolerate, work with, and label sensations in a new way, we use co-regulation and engagement to develop new memories that override previous, challenging ones. It generally takes more experiences of the positive to override the negative memories, but over time progress can be made. *Indeed, it is through our loving and compassionately attuned relationships that our children can learn to navigate the range of positive and negative sensations, thoughts, and emotions.*

A child who can locate and identify what's going on in their own body has a head start on building psychological resilience and, likely, lifelong health. We can educate our children about the importance of paying attention to their bodies and, in doing so, introduce them to optimizing their own mind/body health. In the chapters to come, we will focus on how to nurture this ability further according to a child's age and developmental stage, starting from birth.

Disneyland Lessons

What became of my daughter's refusal to go to Disneyland? We decided to cancel our trip. While my husband and I were excited for the vacation, she didn't accept it as a gift. Instead, we spent her birthday playing and swimming in a neighbor's pool. And she was delighted.

It was several years before I truly understood why she hadn't wanted to go. When I learned more about sensory processing, I came to realize that the idea of a place like Disneyland—which is a veritable cacophony of sights, sounds, smells, and other sensory input—was simply overwhelming to her. She had particular difficulty with *multisensory processing*. In other words, she was highly over-reactive to several sensory systems at once.

When a child interprets the world as unpredictable, they often become more rigid or controlling in an attempt to manage their environment. When children can't predict how they will feel in different environments, life can feel chaotic, and they adapt by being less flexible. My daughter's refusal to go on the trip was her nervous system's brilliant adaptation for her. Her brain and body needed time and experience to figure out how to integrate her senses and to experience them as safe and not threatening.

With this understanding, my husband and I had newfound compassion for our daughter and made sure she knew how much we respected her preferences and choices. Rather than encouraging her to override her body's signals or voicing frustration that she couldn't "go with the flow," we expressed interest and offered validation when she recognized her body's signals. *Indeed, whenever we recognize with empathy that a child's body is having difficulty and provide emotional presence and acceptance, we help enhance the child's ability to tolerate a range of sensations.* Being in strong relationships improves our capacity to stretch and tolerate negative experiences. *When we understand how the body informs the brain, we can help our children create a whole new relationship with their emotions.*

And the story has a coda: Many years later, when she was in middle school, the same daughter asked to celebrate her birthday with two friends—at Disneyland. I still have a photo from that visit: the three of them on one of the park's steepest roller coasters, my daughter jubilantly stretching her hands into the air, her best friends, eyes closed, burrowing their heads into her shoulder. I smile whenever I look at that photo, realizing how much tenacity it took for my daughter to grow through her sensory challenges, and how beautifully her body and brain organized—on her own schedule—along with our emotional support.

RESILIENCE-BUILDING TIP: It's important to understand children's individual differences, including their unique way of taking in information through their sensory systems from the outside world, from inside their bodies, and from other people. Don't assume your child experiences the world the way you do, and be curious about what the world feels like from your child's standpoint. This understanding is the gateway to helping them develop emotional literacy.

7
The First Year

> Every time a parent looks at that baby and says, "Oh, you're so wonderful," that baby just bursts with feeling good.
>
> —T. Berry Brazelton

New parents have so many questions about their infants: How can we get them to sleep through the night? Is it okay to let them cry? Is it okay for babies to watch TV? How important is it to "wear" my baby? Should I teach my baby sign language? Can you spoil a baby?

Kerri and Ben had many of these questions when their pediatrician—who knew of my own struggles with sleep deprivation and stress early in motherhood—referred them to me. Their daughter, Selwyn, was three months old at the time. As we settled in to our first session together, the exhausted parents recounted an easy pregnancy and full-term delivery. They described their wonder at being parents and the excitement of Selwyn's first-time grandparents, making her arrival magical for the entire family.

Yet, far from enjoying parenting, the couple was barely surviving.

At eight weeks Selwyn was nursing well and gaining weight normally. But she cried a lot during the day, and her parents had difficulty figuring out how to soothe her. Feeding her, changing her diaper, taking her for walks, rocking her—nothing seemed to work.

Nights weren't much better: The baby would wake up three or four times, leaving both parents bleary-eyed in the morning.

Over time, Kerri and Ben confided that they had another worry. Their pediatrician had warned them about the dangers of SIDS, sudden infant death syndrome, and listed the precautions to prevent it: making sure the baby slept on her back in proper sleep clothing, keeping her sleeping space clear of bedding and other items. The well-intentioned and necessary warnings were a constant reminder that something terrible could happen to their baby.

It had proved a challenging combination: daytime crying jags, night awakenings, the fear of losing her, and, most of all, their own sleep deprivation. And it all left the couple feeling like they were barely holding on to their sanity. These wonderful parents were doing their best, but they were struggling with the transition to parenthood.

THE FIRST SIX MONTHS

Three of my most vivid memories are of seeing my children for the first time. Gazing into their eyes, I felt something familiar—for I already had developed relationships with them when they were developing inside of me. What I didn't realize before my children were born was how profoundly being a mother would change my life.

Then there's another, less pleasant memory: how I burst into tears when the hospital nurse asked me to change the diaper of my first child, who had been born prematurely. I felt completely incompetent and overwhelmed with the responsibility for the tiny, helpless creature in my arms. How was I to know what she needed if she was crying and couldn't tell me? I hadn't spent much time around babies, so I didn't even know how to change a diaper.

When my husband and I drove home from the hospital with our

little bundle, I dressed her in a preemie outfit so small it might have fit one of the dolls I had as a girl. My love for my daughter and my drive to protect her were the fiercest forces I had ever experienced, a mix of exhilaration and fear.

A visiting nurse dropped by to take my daughter's blood samples and check on her several times daily. Because she had elevated bilirubin enzymes, she had to sleep in a light box, and medical staff advised us to make sure her tiny eye shields stayed on, because staring at the lights could have blinded her. Hardly an easy warning for sleep-deprived new parents to hear! For those first weeks, my husband and I took turns watching her like hawks, day and night, to make sure she kept the goggles on. It was a joyful nightmare.

So I had empathy for Kerri and Ben, as I do for all new parents. Nothing triggers our worry circuits and alters our rhythms and routines like having a newborn. As humans, we love predictability, and when a baby comes into our lives, nothing is predictable. Our routines center on the baby until we settle into patterns around feeding, naps, and nighttime sleep.

And besides sleep deprivation, new parents cope with another challenge: the glut of advice and pressure about how to care for a baby. Of course, nobody can tell you the one "right" way because you and your child are unique. What you can learn is to read your baby's signals and then personalize your strategies and decisions for *your* baby—not some theoretical version of a baby.

Becoming a parent changes our bodies and our brains as, fueled by the power of the love hormone oxytocin, we experience the most significant transformation we have as adults. This is true for both mothers and fathers. Research shows that fathers who were primary caregivers underwent their own hormonal brain changes as a result of actively caring for their babies—evidence of the value of fathers caring for their children as early as possible. Recognizing this, many hospitals and birthing centers now encourage daddy-baby bonding via skin-on-skin contact and cradling the baby just after birth. Parents' naturally surging hormones help them connect with the baby.

In that process, one of our first tasks is finding a shared rhythm, like dancers learning through practice to coordinate their actions with another person. And like any dance with a new partner, it takes practice. Instinctively, we use different tones of voice, facial expressions, touch, and ways of holding and moving together by observing the baby's reactions and adjusting or collaborating to calm the baby and help the newborn adjust to being in the world. Every baby has a unique dance pattern with their mommy and daddy. Sometimes, the dance is easy to figure out, pleasurable, and predictable for each, but sometimes it takes longer to learn—as it did for Ben, Kerri, and baby Selwyn.

Fortunately, research (and my own experience) points to one significant factor that enhances a baby's process of learning to trust the world: loving and attuned parents who respond to the baby's needs. Responding, as I've described throughout the book, means paying attention to a child's body. Our first job is to help babies maintain their body budget. We do this by learning to read the baby's signals.

READING YOUR BABY'S SIGNALS

Babies thrive when their needs are properly met, but they can't tell us what they need in words; we need to figure it out through trial and error. In my training in infant mental health, I learned that the key to supporting healthy infant and toddler development is *responsive parenting*. Responsive parents are warm and affectionate and respond to their babies' needs. They tend to do **three things**: *observe the baby's cues* (e.g., yawning and rubbing their eyes), *accurately interpret the cues* (e.g., guessing that the baby is tired and needs rest), and *take action to meet the child's needs* (e.g., putting the baby down for a nap).

Researchers have found consistently that when parents are responsive to their babies' needs, their infants sleep better, are less fearful, develop healthier eating habits later in life, and experience less stress in their bodies. What's more, these children are likely to have an easier time in preschool, control their impulses and behaviors as early as their development lets them, and be more cooperative. Why? Because the trusting relationships that parents build through predictable, loving, and attuned interactions help a child thrive. *The best thing you can do for your baby is to lovingly observe their cues, interpret what they mean, and then take action to meet those needs.* You might not get it right at first, but go easy on yourself. We figure out a baby's needs through trial and error.

Why is this so important? With our help, infants create meaning through feeling their bodies' internal sensations (interoception) and what's going on outside their bodies (sounds, sights, touch, smell, taste, etc.). They need us to help them physically through our loving connection so that their brains begin to anticipate and make memories of safety and trust. We do this by creating shared meanings together.

CREATING SHARED MEANINGS

We are the architects of our baby's experiences and how the baby figures out the world. Keep in mind that a newborn baby has no memory of the past (outside of the womb). Everything your baby sees, does, and experiences on their own and with you influences how your baby makes sense of the world and the way your baby's brain will make predictions in the future. Parents are the most important force in how babies come to understand themselves and the world.

We learn how to parent each child through trial and error, and it's rarely easy. When I first became a mother, I tried to apply everything I knew about building strong attachments: being present for my baby, letting her know I cared and that she wasn't alone, and trying to comfort her when she cried. What I didn't have were the *specific tools* to comfort a baby who had extreme difficulty settling her body. Theories didn't help me when my baby was screaming and nothing I did seemed to soothe her. I'll never forget the night my daughter cried for three straight hours. I felt so scared and helpless.

Over the years, I've encountered countless parents who, like me that night, have found it difficult to calm their baby, child, or toddler. I always remind them that they're not alone. Remember, when our babies are little, we don't always read them accurately. Growth happens when we try different approaches and learn from and with the baby. This powerful learning happens when we observe and change our approach based on the baby's feedback. The key isn't to be perfect, but to help the baby make sense of the world, knowing that all we say and do is creating memories that help the baby feel that the world is a safe and trustworthy place.

If your child is past the infant phase and you worry that perhaps you didn't provide those experiences, please remember that this book is a no-blame and no-shame zone. And evidence from neuroscience demonstrates why there's no need to panic about how we may or may not have provided responsivity in the past: the brain and body are constantly comparing new experiences with old ones, updating and changing the firing of the brain. While it's natural to regret what you (or others) may or may not have done this morning, last week, or at any time in the past due to the baby's life circumstances or your own struggles, the story of relationships is one of hope. We grow and shift with our experiences, constantly updating our child's predictions about us and the world.

Meeting a baby's basic needs for nutrition, safety, and security is particularly crucial in the first six months of life. A baby's earliest experiences lay the foundation for the brain networks that are to come. Feeding your baby when she's hungry or soothing a child in need of comfort helps build the solid foundation that provides a healthy start. *The way to help a baby move toward self-regulation as a child is first to co-regulate with consistency.*

We provide support for the fussy or crying baby through trial and error. Basically, see what works to help settle the baby. Perhaps it's your caring face and loving voice saying, "It's okay, my love—Mommy's here," and projecting a sense of calmness and trust. Your voice may not be enough to calm them, so they may need to be held *and* rocked in addition to hearing your voice until they settle. Or perhaps a particular rhythm or pattern of rocking helps your baby's body calm. Perhaps you've tried everything, and the baby continues to fuss, so you gently put them down to see if *less* stimulation helps them self-regulate.

Experiment with what works for your particular baby. That's what it means to be responsive: actively interpreting what the baby needs with patience for ourselves and the baby. Research shows that it's a two-way process: Our *heart rhythms* sync with our baby's as we are comforting the baby. It's truly a physical experience of love and building of trust. And let's acknowledge how difficult this can be when we're feeling anxious or upset ourselves—as first-time parents like Kerri and Ben naturally can be.

OBSERVING AND RESPECTING THE BABY'S NATURAL SELF-SOOTHING ABILITY

Once you're confident that your baby's basic needs are met, give them time to discover their body and figure things out with the limited control they have. We can start helping children to build resilience from the beginning by allowing babies to experiment with feeling the boundaries of their own bodies. For example, if you lay down a baby—fed, changed, a bit fussy—in a safe place, they might find their hand and suck on it or stare at an object for a few seconds. Or they might feel their legs push onto the sides of the crib. These moments give babies the opportunity to practice soothing themselves.

Babies aren't completely helpless. They can shift their body position. They can bring their hands near their mouths and even suck on their fists or fingers. They can look around, pay attention to sounds, close their eyes, and sometimes turn their heads away. Of course, different infants have different capacities: Preemies and babies with medical conditions may not yet be able to cry or to instinctively move their bodies. Follow your instincts and offer support when your baby needs support—when you feel your baby has stretched far enough on their own.

RESPONSIVE PARENTING

The three steps of responsive parenting serve as excellent guides: observing the baby's cues, interpreting what they mean, and taking action. *Observing the baby's cues* means paying attention to the signals from the baby's body. *Interpreting the cues* means getting to know your baby and taking care of the obvious *reason* underlying the baby's behaviors (e.g., feeding or changing the baby). And *taking action* means first attending to the obvious needs, then observing the baby for an extra minute or two to see what the baby does next. A baby looking at you might signal that it's a good time for connection. A baby might look away, on the other hand, if they need rest. These are small ways to begin to honor and respond based on the wisdom of your baby's body signals.

A Guide to Being Responsive: Helping Ben and Kerri Cope

No matter our intentions, the sleep deprivation of the baby's early months can compromise parents' platform strength, making everything feel like an emergency. Certainly it felt that way to Ben and Kerri, the couple who called me for help with their infant. When she first called me, Kerri sounded exhausted, her voice sad, flat, and quiet. When they arrived at my office, I could see the toll in their faces, the dark circles under Kerri's eyes. Through tears, she spoke of the joy of being a mother but also of feeling overwhelmed.

The First Months: The Importance of Sleep

The first priority for parents is to establish a minimum baseline of sleep, which, as we learned in Chapter 5, is our life support system. Ben and Kerri needed sleep to think straight, regain their bearings, and find their rhythm as a family of three. Sleep is a nonnegotiable necessity. Yes, it's possible to go days or even weeks with inadequate sleep, but this couple had gone months, and the deficit was taking a serious toll.

To help them recover, we started with a plan for the couple to manage different parts of the night routine, taking turns sleeping until Selwyn developed more predictable sleep patterns and required fewer night feedings. Ben, who tended to go to bed late, covered the first shift, letting Kerri sleep for a few hours. She pumped breast milk so Ben could bottle-feed the baby. Ben also extended his paternity leave by a few weeks so that the couple could take turns napping during the day until they both caught up on sleep. The shift in the parents' platforms was remarkable. Now better rested, their outlook on everything suddenly shifted from a desperate to a more hopeful tone.

Accepting Help from Family and Friends

It's worth seeking out creative solutions to overcome sleep deficits: grab some sleep while the baby naps during the day; enlist friends or relatives to watch the baby while you sleep. In the fog of sleep deprivation, some parents pull back socially because they are conserving energy in survival mode. *In those modes, you may not feel like reaching out to friends or family for help even when you need it.* There's a brain-body reason: We pull back to conserve energy because our stress load is too high. When we are so depleted, we often check out from human contact at precisely the time we need it most. Friends and family might be checking in and asking, "Is there anything I can do?" Rather than turning away help, consider a "yes" attitude, making a list of simple ways that people in your circle can offer relief: dropping off a homemade meal; taking your older children to the park for a few hours; doing a load of laundry; picking up your child from school; visiting for half an hour just to talk.

When my kids were young, a favorite way to recharge my battery was taking a (rare) pajama day. I'd arrange for friends and family to take my kids while I relaxed and stayed home. Since my life was generally scheduled, I'd make no plans for that day and instead do whatever came to me spontaneously: read a book, make myself tea and cookies, watch TV. But the most important part of pajama day was taking a nap—without setting an alarm.

When my friends and family brought the children back at the end of the day, I felt like a new person. I had more patience, and I could delight in my children again rather than simply managing them. It's important for parents of young children to acknowledge and not be ashamed of the toll that parenting can take on our bodies and minds. Remember the parenting mantra from Chapter 5: "This is hard. I'm not alone. May I be gentle with myself." One way we can be gentle on ourselves is by accepting help.

Once both Ben and Kerri found ways to catch up on their sleep, we moved on to the next challenge: Selwyn's crying jags.

Soothing the Crying Baby

Using observation skills, we can read the baby's signals and then use trial and error to see what works to calm the baby's body.

Your Voice

Babies respond to the *tone* of your voice. Calm voices make them feel safe. When you are calm, you will tend to have a more prosodic and musical tone of voice. That's probably why babies love the sound of a parent's gentle voice so much: It's calming because they sense that a person with a naturally prosodic voice is safe. Some dads might feel silly to use a more melodic voice, but it's remarkable how safe it can make many babies feel.

You can also experiment with the *volume* of your voice. Try softening or amplifying it and see if your baby has different reactions. Experiment with varying the pacing, how fast or slow you are speaking or singing. Try different sounds and observe what happens in your baby's body. Does the baby quiet or calm down? Do they look at your face or relax their body? Maybe they close their eyes and turn away. If so, it's a sign your baby needs rest or to try something else.

Selwyn's parents discovered that holding her close to their chests and providing a rhythmic and strong "shushing" sound seem to relax her body during her afternoon crying jags. Every baby has unique sensory preferences.

Background Sounds and Music

Observe your baby's reactions to sounds. If your home has lots of ambient noise, or if you typically have the television or loud music on, see how your baby reacts to fewer sounds. Sometimes, babies need time to just "be" and experience quieter surroundings.

When your baby isn't fussing (or is just starting to fuss), try different forms of background music. Your baby may love Mozart or Disney songs with female voices, or maybe your baby is a rock and roll type. I can't count the number of times I played the Beach Boys' song "Kokomo" because of how effectively it calmed one of my babies and made her smile. Perhaps it was because of its many soothing qualities: a predictable and steady beat, friendly voices, and plenty of repetition. Every baby processes and reacts to sounds (and all sensory stimuli) in their own way. Selwyn's parents, for instance, purchased an inexpensive "white noise" machine and found that she calmed to the sound of rushing water.

Movement

Try different kinds of movement experiences to see how your baby reacts. Experiment with different rhythms and patterns of movement as you hold your baby. Perhaps they like to be held looking out, or snuggled inward, or held like a football with your arm securely supporting the front of their body. See what kinds of gentle movement help calm your fussy or crying baby. Is your baby's body relaxing or becoming tenser when you are moving faster or slower? How about when you rock your baby back and forth or more vertically, up and down?

Selwyn's parents found that she enjoyed being held high on a shoulder, with pressure at a certain point between their chest and her tummy. When she cried, that position seemed to help the most, often producing a burp or two.

Touch and Pressure on the Skin

Just as you can try different vocal tones and types of movement, you can experiment with different types of sleep clothing and swaddles to see how different kinds of touch calm the baby. Sometimes being securely wrapped in a blanket is calming to babies. You can see how the baby responds to gentle arm and leg massages, varying light touch with a bit more pressure.

Selwyn calmed when she was in a firm and cozy swaddle with both her arms inside. In her first four months, she seemed to enjoy having her limbs held securely in—the position that eventually worked best for sleeping.

Removing Stimulation

Sometimes a baby needs less stimulation, not more. When a baby is overstimulated, you might notice them looking away or turning their head away, shutting their eyes, fussing or crying, and—for babies older than three or four months—pushing away if you're holding them. These are signs of fatigue or overload. You can gently decrease parts of the stimulation by continuing to hold the baby, for example, and reducing one activity at a time. If you are rocking the baby while also talking or singing, then stop talking and singing and simply rock. If that doesn't calm the baby, then continue talking and singing but stop rocking and see if there's a difference.

Paying attention to a baby's cry is important because you are laying down pathways of soothing and predictability that babies need in order to develop trust in the world and in their bodies. Still, sometimes babies just need to cry as a way to release energy.

In their first months, babies cry and fuss, and it's our job to figure out if the fuss or cry indicates a type of stress that we can help with—by feeding or soothing, for instance—or if babies just need a little time to explore life in their growing bodies. If you are concerned that your baby is crying too much, always check with your baby's pediatrician or health practitioner to rule out any physical causes that need might need attention.

Ben and Kerri told me that in the afternoons, Selwyn would regularly cry after she had napped, been fed, burped, and had her diaper changed. When the pediatrician checked her out, he said she was going through a colicky period, making the cause of her crying a mystery. (Colic's cause is still unknown, but it appears to be related to the gastrointestinal system, and luckily it usually resolves at around three or four months.)

Fortunately, Selwyn's intense two-hour crying spells stopped by the time she was three and a half months old. In our second month of working together, her parents told me they were starting to feel "sane" again and ready to tackle another major concern: her nighttime sleep.

Nursing or bottle-feeding (if that's your choice or only option) on demand throughout the night and day is one of the first ways we provide responsive care so that the baby's bodily needs are met as soon as possible. In the first few weeks and months, that's how the baby learns to trust the world they're born into and also how we stabilize the baby's physiology. But when you're ready, depending on your baby, your philosophy, and your pediatrician's advice, there are things you can do to encourage nighttime sleep patterns for babies and children of all ages.

Let's Sleep!

There's no shortage of books about how to get your baby or child to sleep. It's best to consult with your pediatrician and other trusted advisers about your specific situation. But it's worth considering some research and tips that can improve sleep hygiene for your whole family.

Ben and Kerri's plan to take turns at night helped them get through Selwyn's early crying spells. But when she turned one, they were ready to try helping her to sleep for longer periods of time. Selwyn had started pushing them away when they tried to soothe at bedtime—one of many cues that she was exploring how to regulate herself. Still, nights were rough since she cried when they left her to sleep, but she pushed them away with her little arms and legs when they offered comfort. How, they wondered, could they help her become sleepy and fall asleep on her own? And how important was it?

As it turns out, sleep is just as important for babies as it is for adults. Sleep researcher Jodi Mindell has found that babies and young children who sleep better have an edge in language development, attachment to others, future literacy, and better behavioral and emotional regulation. Sleep allows all of us to function at our highest capacity.

Sleep Nurturing, Not Sleep Training

One of the most common questions people ask new parents is: "How's the baby sleeping?" A great deal of parenting education focuses on sleep, yet I avoid the term *sleep training*, since it implies there's a specific way to train. There's no one "right way," since no two babies are alike. We need to see what works for each child and modify it as the child grows and faces new sleep challenges. Of course, when it comes to sleep, parents have a broad range of choices, preferences, and practices. Some parents co-sleep with their children—either by choice or necessity. It's best to follow the advice of your child's trusted pediatrician. The American Academy of Pediatrics recommends that "infants sleep in the parents' room, close to the parents' bed, but on a separate surface designed for infants, ideally for the first year of life, but at least for the first six months."

Some people associate sleep training with letting a baby cry for long periods of time. In my opinion, that's neither desirable nor necessary. Instead of training, I use the term *sleep nurturing*. As a parent, you don't have to shift from being emotionally attuned during the day to a trainer at night. Sleep nurturing, in fact, can be responsive and consistent with attuning to your baby's emotional needs during the day.

When we shift our mindset to *nurturing* a child rather than *training* a child to sleep, the process becomes a natural outgrowth of responsive parenting, with the goal of helping babies and children look forward to sleep as something that is *relaxing, predictable, and safe.*

Sleep presents children with several challenges. First, if the child has developed a habit of falling asleep only with your help, you might want to consider shifting that pattern. Otherwise, every time the child wakes up (as we all do after cycles of 90 to 110 minutes of sleep) they'll need help falling back to sleep. One of Selwyn's challenges was that waking up was a signal that she needed her parents to help her settle back down.

Second, sleep involves shifting into a different state than being awake, so even if your child sleeps with you, the child experiences a *separation*. Sleeping means losing the ability to keep track of you and of consciousness itself. If your child has their own sleep space, the separation is physical as well. *It's understandable, then, that sleep would have an emotional component, and we should expect that it's a developmental process that takes time for each child to sort out.* In short, it's a big step, and patience and attunement are key. As with everything else, we need to help our children work in their just-right challenge zone that is respectful of their body budget and platform.

Sleep as a Predictable, Enjoyable, and Relational Experience

Research demonstrates that one key to helping children get to sleep are bedtime routines, "the predictable activities that occur in the hour or so before lights out, and before the child falls asleep." Babies and children with predictable bedtime routines sleep better—and, not surprisingly, so do their parents.

Routines in the hour before sleep can provide a soothing, relaxed experience, something positive for children to associate with sleep. Bedtime rituals are comforting since they offer predictability, which helps children of all ages feel safe. Of course it's also useful to consider timing of naps and bedtime to ensure that your baby or child is sufficiently tired and ready to sleep.

The routines that Dr. Mindell and her colleagues found help young children sleep include *four basic components* in the hour or so prior to putting the child to sleep:

Nutrition (e.g., feeding or a healthy snack)
Hygiene (e.g., bathing or brushing teeth)
Communication (e.g., singing lullabies, quiet talking, reading a bedtime story)
Physical contact (e.g., cuddling, holding, massage)

These categories aren't meant as rigid requirements but rather as general guides to customize to your child's preferences, your parenting style, and your family's schedule and preferences. If a baby has sensitivity to water on the skin, for example, then a bath might not be a good choice for the bedtime routine. And some babies enjoy cuddling and massage, but others prefer only a light touch or simply to be rocked. Some babies might only need one or two of the activities on the list. See what works for you and your baby or older child, based on what you discovered about their sensory soothing preferences in Chapter 6.

Just after her first birthday, Ben and Kerri were eager to devise a plan to help Selwyn sleep longer at night. We discussed the benefits of a more predictable nighttime routine that involved putting her in her crib sleepy—but not fully asleep—to see if Selwyn could be nurtured into sleeping for more of the night. They settled on a routine that incorporated aspects of each of the four bedtime categories. They fed Selwyn her dinner (now including some solid foods) an hour before bedtime. Next, they gave her a bath, followed by nursing while she was still alert and awake, rather than nursing her to sleep. Then they quietly played with her for about ten minutes, softly rubbed her back, closed the blinds, sang a short good-night song, kissed her good night, and put her down. Selwyn enjoyed those steps, and they felt comfortable to each parent, creating a slow transition to bedtime. Again, each child is different, and your baby or child may not require as many steps as Selwyn did.

Over the next several weeks, her parents went in and quietly soothed her when they felt she needed support. Eventually, when she woke up at night, she fussed for a few minutes at a stretch, then fell back to sleep on her own. Within a month, Selwyn was waking just once a night to nurse. This proved to be a workable pattern for Kerri and Selwyn. Both were waking up only once a night, and Kerri loved the closeness and intimacy of the silent night feeding. Several months later, Selwyn stopped waking up in the middle of the night to feed. Now the whole family was sleeping through the night, except for the occasional and expected disruptions, such as when Selwyn caught her first cold and woke up from a stuffy nose, and when she was uncomfortable from teething.

No matter the age of your child, soothing bedtime routines provide a wonderful time to connect, slow down, and create predictability. The four parts of the sleep routine are also basics of healthy child development, encouraging good habits throughout childhood and beyond. These healthy activities—a nutritious meal shared with parents, unrushed communication through dinner and bath time, physical and loving contact—are all big deposits into a child's body budget. We use our relationship as the *lead-up* to sleep as a predictable, enjoyable, and relational experience.

If your best attempts at helping your baby sleep still don't work, remember that some infants simply have an easier time figuring out the sleep-wake cycle than others. Stay hopeful, try new things, have compassion for yourself, and surround your baby with cues of safety to their nervous system. And reach out for help if you feel overwhelmed. Your child's pediatrician or an infant-sleep specialist can provide customized support if necessary.

As your child grows, the nighttime routines change with your child's evolving needs, but it's always nice to end the day with a predictable moment of connection. With everyone in the family sleeping better, you can derive more enjoyment from the burst of development that happens every day in the baby's first year—when communication and spontaneous exploration and play happen all day long.

The Second Half of the First Year

In a baby's first months, we often feel astonished by how quickly the baby is growing: "They're so big!" "Did they grow taller overnight?" Then it seems too quick. "Slow down," parents say. "I want to enjoy this stage for a few more days!" The rate of growth in the first year is remarkable, as the brain develops a million connections between neurons every second. From the moment of birth, the newborn, staring at your face and reclining comfortably in your arms, is experiencing a developmental explosion. Every day is full of growth and exploration.

In the second part of the first year, babies develop more and more control, and the world becomes one big science experiment: reaching for objects, mouthing them, looking at them, dropping them, pulling things out of drawers or boxes and putting them back in. Each day, babies gain a bit more body control, first staring at a mobile, then batting at it with uncontrolled swipes, then reaching more purposefully for particular objects.

Imagine the powerful drive to gain control, little by little, over your body—and the exhilaration that comes with it. Suddenly you realize you can move your arm and swat at something you have been looking at! That control expands to grabbing things and putting them in your mouth. In the first year the baby goes from gazing at things to exploring the world with their whole body. With your help, your baby is learning to enjoy and tolerate a world of new sensory experiences.

Every skill a baby acquires moves the child further into the world of exploration: sitting up, crawling, cruising, and eventually walking. Interacting with the world is alluring and fun for your baby. They want to manipulate, hear, taste, tear, move, and drop things! Everything is new and they are experimenting with cause and effect.

And babies are drawn to repetition. They find satisfaction and actively learn about patterns by doing things over and over again. Have patience, keeping in mind that they are scientists and must discover everything anew by *doing it, rather than watching*. Your job is to keep the baby safe—and doing so requires vigilance.

You may be tempted to say "no" frequently in the first year but remember the balance between nurturing a baby's natural drive to explore and setting limits too soon. Babies aren't yet aware of safety—it's a learning process. It's healthy for babies to want to figure out how the world and their own bodies work. As we introduce necessary limits to help the baby stay safe, we also want to nurture that incredible drive to explore and try things—of course, in ways that make sense for you and your family's situation.

COMMUNICATING THROUGH PLAY

Three simple words guide our interactions with babies and children of all ages as they engage in the most powerful development booster of all: play. Follow their lead.

Start with simply being present, setting aside phones and screens, and focus on being curious. Since babies are exploration machines, we don't need to orchestrate much. If you follow your baby's lead, your baby will show you what they are interested in. The baby's natural curiosity is a driving force that, as your baby progresses through the different stages of the first year, leads to the natural desire to connect and communicate with you through the all-important serve-and-return process.

Think of play in three broad categories: **self-directed exploratory play, play with a parent or other adult, and play with peers.** In the first six months of life, most of a child's play is the first two kinds: looking at, touching, manipulating, or mouthing objects, and eventually sharing that experience with you. These are important ways to explore the world. Play with peers doesn't come until later.

When you follow your baby's lead, eventually you'll find that your baby wants to do something magical with the objects they're exploring: show them to you or hand them to you. This simple interaction is the basic building block of communication, an extension of the serve-and-return, which began as tender eyes gaze back and forth, as your baby shares their world with you.

They'll grab a rattle and hand it to you, expecting a response. What will you say? Your baby might bang a pot with a wooden spoon and then become startled at the sound, looking at you for reassurance. "Oh my, that was loud!" you might say. "It's okay, sweetie." They may feed themselves and then if you open your mouth, they may put food in your mouth, their first time sharing something, as you light up and say, "How delicious! Thank you!" Your baby smiles and then tries to feed you again. This is play—what happens when we follow the baby's lead. The simple back-and-forth exchanges of laughter and cooing in the first six months eventually morph into sharing objects with you in the second half of the first year, and eventually to the baby gesturing toward what they want—a huge step in your baby's ability to communicate with you. Pointing is a precursor to social problem-solving, and following your baby's lead in play is a powerful building block that sets up the ability to solve problems throughout life.

As the serve-and-return grows, something wonderful happens: The back-and-forth grows more sophisticated, and the play becomes an exercise in platform strengthening, allowing baby and toddler to experiment with new ways to move and experience a wider range of feelings. That's why play is called a *neural or brain exercise*. In fact, it's the best brain-body exercise, helping a child build self-regulation while integrating a range of sensations, ideas, and emotions in the fun and appealing context of relational safety. A perfect example: the game of peek-a-boo.

The Magic of Peek-a-Boo

Playing peek-a-boo with a baby provides an exercise in holding a moment of stress or tension (when the playmate disappears) and then releasing it (when the playmate reappears), to the baby's delight. A peek-a-boo game is a just-right challenge for babies turning into toddlers, as they learn and confirm, over and over again, that when someone disappears from view, they still exist.

Later in childhood, more complex games such as hide-and-seek provide the same kind of tolerable tension—tension that is relieved when you find the person hiding or they find you. These play activities support a host of emerging problem-solving skills, while providing fun at the same time. They prepare children for their growing independence in enjoyable and pleasurable ways. We'll explore play with toddlers and older children in subsequent chapters.

SELWYN'S LAUNCH

Just before Selwyn's first birthday, her parents paid a visit. Cautious at first, she held her mom's hand as they entered my office, where we all sat on the floor and she grabbed a toy from a bin, showing it to her dad. Scooting aside to enjoy watching the young family play together, I recalled the parents' haggard looks of six months earlier, a contrast with their smiles now, reflecting the solid platforms they had since built. Then came the added surprise: Selwyn stood up and started walking and wobbling across the room. Just that week she had taken her first steps. When she reached the end of the room that led to a hallway, she looked back to her parents with a big smile on her face, a sign of the trust and safety they had all built, as she started toddling ahead of them into the big playroom.

Watching her reminded me how important the first year is for building a strong brain-body connection. When we meet infants' needs, they feel safe and begin to have trust in the world, and use that trust to reach out, communicate, and explore. This is the solid foundation that leads to the ability, one day, to share their ideas, to help you understand what they need, and to solve problems together with you.

RESILIENCE-BUILDING TIP: Responding to your baby's needs involves observing the child's body language, figuring out what it means, and then taking loving and attuned action to meet that need, while respecting the baby's own growing abilities to self-soothe. It may be difficult, but turn down the noise of the parenting advice that goes against your gut and rely on those instincts that lead to being responsive to your unique and precious baby.

8
Tantrums Throw Toddlers
PUTTING TODDLERHOOD IN CONTEXT

> Calmly support children through their frustration, disappointment and even failure, so that we normalize these difficult but healthy life experiences.
>
> —Janet Lansbury

Life with a toddler can be a wild ride. Understanding toddlers' emotions and behaviors in the context of the developmental process can make it more manageable, helping us to appreciate and celebrate their natural curiosity, creativity, boldness, and exuberance in exploring the world.

I was at the airport in a slow-moving security queue when I noticed, just ahead of me, a couple with a toddler who was getting increasingly frustrated. He began to whine and his parents gently tried to calm him. But as we all snaked through the line, his whine grew into a loud request for his Fluffy Bear, which had apparently been mistakenly packed in a checked bag. By the time the parents approached the TSA agent to show their IDs, the plea had escalated to a demand and the boy had thrown himself on the floor.

Watching this familiar scene, I felt my heart beat a little faster, my mirror neurons firing. I watched the stressed parents doing their best, gathering their son from the floor as he kicked and screamed. Excusing herself from the TSA agent, the mother stepped back and waved me ahead. I passed, listening to the parents conferring aloud. "Pick him up?" the dad asked. The mom shook her head. "Let's just wait," she said. They moved aside, trying to keep the boy safe as he thrashed about. Moving forward toward the terminal, I could still hear his wail, and I thought back to how difficult it had been to travel with my own toddlers.

A few minutes later in the terminal, I spotted the same family, the dad holding the little boy, now asleep on his shoulder. I smiled, thinking of the many parents of toddlers I've worked with over the years, so many of whom share the same struggles: How to manage a toddler's rapid growth, shifting emotions and behaviors, and, of course, tantrums?

I thought of Roger and his husband, Vince, a couple who had recently met with to me to discuss their son, Jordan, adopted at age two through the foster care system. By the time he was four, Jordan's preschool had suspended him twice. Then the school's director asked me to help. Jordan had recently shoved another child after an argument over some toys in the sandbox, and now his parents were concerned about his ability to manage his anger. The director told me that teachers had tried everything they could think of to help Jordan "use his words" and not his actions, but to no avail. Roger and Vince were confused, too. Sometimes, they could easily reason with Jordan when he was calm, but other times, he seemed to lose control over something as small as one of his parents accidentally pouring milk in the sippy cup usually meant for water.

Toddlers are often unpredictable, and their behavior leaves many parents feeling confused. Here are a few of the challenges familiar to many parents of toddlers:

Tantrums

Power struggles

Being disagreeable

Whining

Shrieking

Saying no to everything

Testing limits

Not understanding that "fair" doesn't mean getting everything they want

Asking for something and then not wanting it once they get it

Constant negotiating

Not knowing or being able to say what they want or need

Biting

Hitting

Pulling hair or scratching

Throwing food or objects

Sibling rivalry

Consider a few stories my patients have shared with me:

"My toddler begged and *begged* for toast and butter. I finally gave in and made the toast, handed it to my toddler, and he burst into sobs. He screamed 'NO TOAST!' and hurled the toast on the floor."

"After a wonderful and fun day at the zoo, my three-year-old asked for another one of the dinosaur candies I bought him at the gift shop. I let the child know that he couldn't have the candy because it was almost dinnertime. He begged and begged and when I didn't relent, he screamed for twenty minutes in the car until we got home. My ears were ringing."

"In the middle of a meltdown, I asked my child what might help her to calm down. She didn't answer, so I offered a few suggestions.

She said, '*No!* I don't want to do that!' So I became quiet and stayed calm, and she then immediately demanded, 'Tell me what to do!'"

"My son went through a phase of being incredibly difficult to please. It wasn't tiredness or hunger. I felt like all his needs were met, but he would just lose it and I would have no idea how to get him out of it. I would often ask if he needed a hug. Sometimes he said yes, and there were times I was hugging him with gritted teeth wondering what the heck his problem had been moments before!"

What's a parent to do? Ignore the behaviors you want to discourage? Pay attention to what you want your child to do more of?

Of course, there isn't a simple answer, and a variety of books, podcasts, blogs, and other sources offer an array of opinions. Some say to approach tantrums with love and understanding, providing empathic support while enforcing boundaries and also allowing the child to express their negative emotions. Others suggest using time-outs, lest the child learn that a tantrum is the best way to get what they want. Some experts embrace a cognitive approach: patiently trying to reason with the child and teach the child about their behaviors and what they mean. Others take a less active approach, remain calm, and wait it out; tantrums are inevitable. You might well hear different advice from your pediatrician, your child's preschool teacher, your friends, parents, and even strangers in the checkout line at the grocery store who want to chime in while your child dissolves into a fit of tears.

In this chapter, I'll explain how we apply everything we've learned so far to toddlers. We will discover that customizing your approach to your child's nervous system is the key strategy for supporting the ups and downs of toddlerhood. We use the platform, and not the label we assign to the behavior, as our guide. Our guide should be the child's nervous system: If a child is on the red pathway, we move in one direction; if not, in another.

How can you know what's best for *your* toddler? Let's start by shifting the conversation from how to manage behaviors to understanding more about toddlers' capacity for self-control.

A PRIMER ON TODDLER DEVELOPMENT

Children progress rapidly from infancy, when they have very little control over their bodies, to toddlerhood, when they are able to point and otherwise use their gestures—and eventually their words—to communicate. This remarkable developmental explosion happens quickly, and the growth spurts can be deceptive. Toddlers can seem so grown-up: asking, showing, and telling us about their world. But that doesn't mean they have the ability to control their behaviors reliably, see the bigger picture, plan ahead, or make good decisions. Those abilities, or *emerging skills*, are very much still under construction; we can't expect that a child can behave this way all the time. As we learned in Chapter 1, learning to think ahead and reliably control one's emotions and behaviors (our executive abilities) is a developmental project that *begins* to emerge in toddlerhood and continues all the way through the mid-twenties. In short, we often expect more from toddlers than they can deliver.

Neuroscience offers one explanation for why a toddler might melt down if the ketchup meant for the chicken nuggets gets mixed in with the peas on their plate. It's related to something called statistical learning, which is basically the brain's ability to learn patterns. Toddlers are early in the process of gathering enough life experience to build a large repertoire, so they're still learning to make accurate predictions about their environment. They are also very early in the process of understanding that other people may have different thoughts and opinions from theirs. *(You mean you didn't know that I don't like the peas touching the ketchup?!)* This is why a toddler can regularly become dysregulated for seemingly bizarre and small reasons: It's understandably stressful when reality doesn't comport with your expectations. A child's response to such disappointments depends on how strong their platform is in the moment—in other words, the balance in their body budget. I hope this bit of neuroscience helps you understand why tantrums are so common (and expected) in toddlerhood.

All of this leads to the expectation gap, which we discussed in Chapter 4, the misunderstanding that happens when we assume that children are capable of doing things that their brains and bodies simply aren't ready to do yet. That gap peaks in toddlerhood. We think these children can control their emotions, request things politely, curb the urge to hit or kick, and follow directions. *But toddlers are just getting on the road to acquiring these abilities.* This is why we need to have compassion for them (and for ourselves) as they learn how to predict more effectively, adopt differing perspectives, and be more flexible mentally. It takes time and it's a wild ride, but when everything is in sync, you get glimpses of the mature person your child will one day be. And the next thing you know, they're throwing toast in your face!

It's our unreasonable expectations of tots that explain phrases such as "the terrible twos" and "tantrummy toddlers." *But instead of viewing toddlers' behaviors as "bad" or undesirable, we should see them as expected markers of robust development.* They happen for a reason: It's a toddler's job to explore and figure out how the world works while their capacity for self-control is in the earliest stages of development.

Indeed this is the kind of behavior we should *expect* from toddlers: asking for cake instead of cereal for breakfast; wanting to draw on the walls with lipstick; trying to pull down a heavy rake hanging in the garage; grabbing things from siblings instead of asking politely; resisting being pulled from a favorite activity. It's all part of exploring their world!

Many of the parents and teachers I work with *overestimate* toddlers' abilities to control their emotions and behaviors. That explains why we have such high rates of suspensions and expulsions of little ones from preschool in the United States. We need to start by asking, Is the behavior top-down, an attempt to test out an idea or hypothesis? Or is it bottom-up, an indicator that the child's body budget is depleted and the toddler has exceeded the just-right challenge and moved into the too-much challenge zone? Remember that toddlers don't decide that a challenge is too great; their bodies make that determination for them. *When they move into the too-much challenge zone, they need us to make a deposit into their body budget—with our emotional attunement.* We need to understand that their young nervous systems are struggling to feel calm, and their behaviors reflect their early stage of development.

None of this is easy. Parenting toddlers is physically and emotionally exhausting and unrelenting—and the stakes are high. Toddlers don't yet know how to keep themselves safe. They need us to teach them to be safe—to continually set appropriate limits and help them learn what they can and can't do.

When we understand that toddlers don't yet have the ability to act like the polite citizens we want them to be, it shifts our expectations and helps reduce our own frustration. It also helps to remember that toddlers gain self-control and safety awareness on their own schedules. It's not helpful to compare your toddler to their peers (or siblings, for that matter), because each child is different and developmental timetables vary. You will certainly see other toddlers doing things your child can't do (or vice versa), but try to resist falling prey to the expectation gap.

When we have unrealistic expectations of toddlers, we risk

sending them messages that if they can't control themselves, then they are *choosing* to do something "wrong" and being "bad." Our tone of voice or the look on our face instantly sends a message of approval or disapproval that the child absorbs. Consider these statements: "If you don't calm down, we are leaving this playdate now!" or "If you don't stop crying, no screen time tonight!" While these may sound like viable incentives for someone who is *in control of their emotions*, a mid-tantrum toddler can't process them. Instead, they pick up on the lack of attunement, which increases the threat level so that the child becomes *more* rather than less upset.

Toddlers are subject to feeling embarrassment and shame, just as we all are. When we correct a toddler with a negatively toned statement when they are out of control, we're using shame unproductively, implying that the child *chose* to be out of control, when the child didn't. This affects the child's self-image. Of course there are exceptions: If a child runs into a busy street, we might yell or grab the child to keep the child safe. But we can use a softer approach for harmless and safe exploration, *and* when a child has lost control over their emotions. We should respect the child's natural and good-natured sense of curiosity and exploration, knowing that it takes children time to understand the impact of their behaviors fully.

In short, we don't want to blame a child for something that's not yet in the child's control. *Toddlers don't generally throw tantrums, tantrums throw toddlers.* We need to understand that toddlers' emotional growing pains aren't simply "negative" or "attention-seeking" behaviors. In fact, we *want* children to express a range of negative and positive emotions because these variations of experience are a robust part of being human and reflect the nervous system's ability to react to changes. Of course, we give children clear, loving, and firm boundaries and expectations—children need them. But we shouldn't punish children for stress behaviors.

TANTRUMS VERSUS LIMIT-TESTING BEHAVIORS

During this stage of minimal self-control, the child is also learning about the world, about what's okay and what's not okay. Naturally, that process includes testing limits. Think about all the things that toddlers have done *to* them, and all the decisions that are made *for* them. This is happening just as their ideas and ability to explore are blossoming. They crave control in their little worlds. I think the word *exploration* is a better term to describe toddler behavior than *limit-testing*, which has a negative connotation. Toddlers are discovering the power of connecting their ideas with their abilities and discovering the limits of their autonomy. That's not easy! On the one hand, you feel like the world is your oyster, and on the other hand you discover that you can't do everything you're imagining or thinking about. That's why we're best being patient, relaxed, and even-keeled in our teaching. Exploration is a part of every child's natural development. *We should meet it with emotional acceptance of a child's negative and positive emotions while maintaining firm and clear boundaries and teaching when it's necessary.*

Consider two examples:

LIMIT-TESTING: You see your child sneak into the fridge and grab some cookie dough to eat. The child's body is calm, their actions deliberate and planned—actually brilliant! But here's where our knowing better comes in. First of all, it's a no-no, and secondly, it's not safe since it has raw eggs. This is the time to teach patiently and set a boundary.

TANTRUM: Your child melts down when it's time to leave the birthday party. How do we know it's not an intentional act of defiance? The child shows signs of the red pathway's fight-or-flight response: red face, eyes darting, lots of movement, yelling or stress in the voice, perhaps kicking and running away from you as well. (Fun, huh?) Time to calm the child's platform.

Toddlers have a *long* way to go toward controlling their emotions and behaviors. Children develop that kind of self-control through co-regulation. Over many years, a steady relationship of calm presence, attunement, and clear and consistent boundaries leads to improved self-control, as the child learns how to use their thoughts to manage feelings and emotions.

PARENTS UNDER THE SPOTLIGHT

I meet many wonderful parents who wonder if their toddlers' "naughty" behaviors say something negative about their child or the quality of their parenting. It's natural to feel that our children's behaviors reflect on our parenting skills. But the truth is that what we consider misbehaviors, especially in toddlerhood, are an expected and natural part of development. Toddlers grow through their abilities to test cause and effect. Our job is to keep them safe while they experience that natural drive to explore. No amount of excellent parenting can prevent certain toddlers' occasional tantrums.

To counter the expectation gap, we can realign our expectations and solutions associated with the power struggles, which can take a toll on our budding relationship with the child. To develop the brain pathways to control their emotions, toddlers need thousands of hours of responsive care (see Chapter 7) and co-regulation (see Chapter 4). This is how they figure out over time how to calm themselves, or to find help in order to regulate or manage big emotional challenges.

Consequently, we have to consistently set firm, loving, and unambiguous limits. But that doesn't mean that we can't try to understand the child's perspective, intentions, and feelings *about* those limits. After all, the ability to make a case for eating cake instead of healthy cereal for breakfast is a sign of early debate skills! So is explaining why you want to go to the park instead of staying inside—an early exercise in self-advocacy. This is good stuff. We can appreciate a good debate—and the range of negative feelings, including anger and frustration when we thwart the child's ideas—while still holding the line of what we know is best for our child. They aren't incompatible, as we will see.

SOLUTIONS FOR MANAGING TODDLERS' EMOTIONS AND BEHAVIORS

Remembering the Difference between Top-Down and Bottom-Up Behaviors

Let's review one of the most important questions you can ask about a child's behaviors: Is my toddler's behavior in this moment top-down or bottom-up? Different behaviors call for different solutions, and the answer to this question will determine what to do next and whether you should start by reasoning or co-regulating. There will be times for both, but the first step is understanding the difference between *top-down* and *bottom-up* behaviors. First, let's discuss bottom-up behaviors, like tantrums.

BOTTOM-UP BEHAVIORS are instinctive and result from a child's body budget running low or being in a deficit and/or their safety-detection system registering levels of challenge or threat. We call them bottom-up because they come from cues in the body and involve behaviors that are driven by their physiology rather than willfulness or limit-testing.

Remember that the child's safety-detection system can signal danger even when the child is objectively safe. It's so subjective. Sometimes, to a toddler, having their banana cut up into four pieces instead of two feels like a threat! Of course, that defies logic, and that's why it can be so difficult for us to be gentle and empathic. *But remember that toddlers are still developing the ability to figure out whether something is rational or not and to tell themselves that everything's okay. They need our help to do that.*

They are also still learning how to modulate or control the intensity of their reactions to situations. *Modulating* means regulating or tempering according to measure or proportion. Eventually, children develop the ability to use their words and top-down thinking and react in a way other than crying, yelling, or otherwise melting down when they experience disappointment. Over time, a child develops the ability to think, *Maybe it's okay if my banana is cut into two pieces because it will still taste the same*, or even to ask a parent for a (dull) knife to slice the banana themselves. These adjustments may seem simple, but they are actually complex top-down reasoning abilities that most toddlers are still developing.

Logical thinking, self-talk, self-regulation—toddlers don't yet consistently have these capacities that help us manage disappointment or unexpected events later in life. That's why toddler behaviors can be so trying and seem so illogical. And depending on the balance in the child's body budget, a toddler's behaviors can spiral out of control in an instant.

That child I saw at the airport, for instance, couldn't ask his parents for help or talk himself off the ledge. He was in a *bottom-up* situation. His parents were trying to wait out the disruption in order to get through the awkward situation. It didn't appear that they had any strategies other than to let him work through it. That's one way to do it, but it helps to have more proactive strategies.

Jordan, the preschooler who would suddenly shove a playmate or grab a toy, was also dealing with bottom-up behaviors. His platform was reacting to sudden changes in his safety-detection system that registered deeply unpleasant feelings in his body. And a history of trauma in his first two years of life made Jordan more prone to misreading social cues or registering objectively safe situations as a threat. (It's also worth noting that Jordan is Black, making him far more likely to be targeted for punishment for his behavior problems than his white counterparts due to the implicit bias and racism that plagues the educational system in the United States.)

TOP-DOWN BEHAVIORS are deliberate and intentional. We call them "top-down" because they are linked to the top part of our bodies, the brain's executive functions. Remember the attributes of successful executives, including the ability to adjust and flex with change, consider the big picture and the details at hand, and make decisions based on past learning and anticipated results. Toddlers certainly have *emerging* top-down abilities and behaviors, but we can't expect them to perform at adult levels.

TODDLERS' SELF-CONTROL IS SPOTTY AND EMERGING

Your toddler may sometimes appear to be in control, but then lose it quickly. Out-of-control behaviors—screaming, crying, hitting, kicking, running, throwing things—are most often instinctual or body-up behaviors indicating activation of the fight-or-flight response of the sympathetic nervous system. Behavioral challenges, if accompanied by the activation of the red pathway (the fight-or-flight response), are an example of bottom-up (or body-up) behaviors. These behaviors are responses to the child's safety-detection system being activated. In those moments, the toddler is surviving, not thinking. Toddlers don't yet have the full ability to calm themselves or use logic, planning, or reasoning to figure out a plan to self-regulate and stay calm. That ability takes many years to develop and many adults struggle with it as well.

ACTIVITY: Try to remember the last time you had an "adult tantrum" and lost control over your behaviors or emotions. If you were with another adult, did that person try to reason with you or convince you to change your mind? If so, was that helpful? What if that person had abruptly turned away to ignore you—or just walked away? Or maybe you were with someone who saw your struggle, understood, and realized that what you needed was presence and acceptance, someone to witness your pain and hold you (either in mind or body). Most people, no matter how old, want to be seen and witnessed without judgment.

DON'T RUSH TO TEACH TODDLERS HOW THEIR BRAINS WORK

With our growing awareness of neuroscience, it's natural to want to teach our children about their own brains and pathways. But we build self-regulation and self-control first through relating, not teaching. We need to wait to teach toddlers about using their brains to control themselves until they're developmentally ready to do so. *Before we teach self-regulation strategies, we need to make sure the child has the requisite groundwork and foundation of co-regulation.* Otherwise, it's frustrating for child and parent alike.

If you have tried to teach your child—of any age—strategies to help their own behaviors and it's not working, then it's probably too soon; you may be asking them to do something that they are not yet developmentally ready to do. Many preschools have posters and charts matching facial expressions with emotions to teach children how to label feelings and calm themselves down. These are neither scientifically grounded nor culturally sensitive. Contrary to popular belief, facial expressions do not universally match up with emotion states. Some schools have "time-in" corners where children can retreat to calm themselves. Of course, everyone loves a cozy corner now and then. But what truly helps children build brain-body pathways to self-regulate is having the experience of loving attunement with caring, patient adults, and not matching pictures of faces to words.

We have to do the necessary work of *relating* before we *teach*. Psychological resilience is built primarily through relationships, and *not* through teaching children how to behave or even teaching children (especially toddlers) how to calm their bodies down on their own. Our early education systems need to understand this, because too many use teaching models that hold children responsible for self-regulation too soon. *When we forsake co-regulation in favor of rules, discipline, and group values (as in situations where care is guided by a chart rather than children's bodily cues), we can weaken the platform a child needs to learn self-control.* This is the difference between teaching toddlers *about* self-regulation and building their platforms *for* self-regulation.

But there's plenty we can do. We can nurture emotional literacy early on by noticing and helping children take note of their bodies' sensations. We can then help them find simple words to label their sensations. *This is a human superpower: using words and concepts to understand our experiences and share them with others.* Before we turn to the solutions for top-down behaviors, let's first look at solutions for bottom-up behaviors.

SOLUTIONS FOR BODY-UP (BOTTOM-UP) TODDLER BEHAVIORS

When we know a child is reacting instinctively in a bottom-up mode, then we use bottom-up solutions. In other words, if a child's body is in the throes of a tantrum (a classic bottom-up behavior), then the first step is to try calming the body, if that's possible. Behavior shifts are often signals of a body-budget deficit and signs that the child needs emotional support, different pacing, or fewer demands to get back to the child's appropriate challenge zone. What they *don't* need is immediate ignoring or strong discipline.

If your child's behavior is a bottom-up stress reaction, use the check-in (see Chapter 3): check your platform first, then check the child's platform. Then use body-up strategies to help the child's body come back to a state of calm so they can engage in connection with you again.

The Simple Formula for Helping Kids Find Calm When They Are Upset: Resonate and Respond

Your child suddenly begins to cry because they wanted the colorful sprinkles on their cupcake, but there were only plain ones left to choose from. Toddler behaviors aren't always logical or proportionate. But a simple two-part formula can help calm emotional reactivity for children—or adults, for that matter. In fact, these two steps can help with many common power struggles between parents and children, or even between adults.

When the tantrum hits, stay close and first observe for a few seconds to see if the child is trying to find a way to figure out the problem or calm themselves. If your child is downshifting on their own, then marvel at the emerging self-regulation and offer your gentle presence—and help, if asked. But if your child is *not* calming on their own and is distressed, angry, agitated, anxious, checking out, or otherwise in need of emotional support, try these two steps: First **resonate,** then **respond.**

RESONATE: Begin with the response that reaches the child's nervous system most quickly. When highly agitated, as we've learned, a child can't hear your words well—they're poised to run, not listen. *So start resonating by offering a form of nonverbal recognition with your emotional tone, your voice, and/or your body language.* This communicates to the child that you *see* their distress and that you are *emotionally and nonjudgmentally present* with them. As the word implies, we metaphorically reverberate or vibrate with a similar energy or emotional level as the child. We share our regulation through our calm presence.

When we resonate, we try to be nonjudgmentally present, and we *adjust our body position and posture* in relationship to the child, perhaps sitting near the child or kneeling close by. Then, *adjust your eye gaze* to what's comfortable for the child (direct, peripheral, or no direct eye contact, depending on what is comforting) and focus on the *tone* of your voice, making it gentle, tailoring it to your child's preferences as we discussed in Chapter 6, and watching to see if your child softens or relaxes a bit. You might add a physical component, such as a hand on the shoulder or soft touch of the arm, but only if your child calms with touch, in the moment. *Sometimes the best possible way to co-regulate is through supportive silence, to be lovingly present without words.*

It may feel awkward and take some experimenting with different approaches to discover what works for your child: a gentle sigh while gazing with a compassionate or understanding look; momentarily mirroring the look you see on the child's face, so they feel understood and less alone. We are so inclined to manage and problem-solve with our words, and not to resonate without automatically talking. With this first step, your child experiences what it feels like for someone else to hold them in mind. The child feels less alone in their distress when you communicate that you accept their range of all emotions. This provides a powerful building block for children's future trust in themselves. Deb Dana, a psychotherapist, says this deeply attuned co-regulation provides a *tether to the grounded state of the nervous system.*

This is how we begin to help a child recover from a tantrum. By sharing our pathway in this way, we help the child find their way back to a feeling of safety. One caveat: If *we are triggered ourselves*—not uncommon during a child's tantrum—it's difficult to resonate successfully. This is such an important point that I hope is clear by now: We can't co-regulate very effectively when our own nervous system isn't grounded. As we learned in previous chapters, we share the state of our nervous system with those around us, whether we like it or not. And that's also why self-compassion and the other techniques we learned about in Chapter 5 are so important.

While we resonate, of course, we might spontaneously use words that capture the emotions of the moment: "Oh, my sweet child," perhaps, or "Ahh, this is tough," or some other short, gentle statement that conveys acceptance and not criticism. *But the essence of the resonating is the nonverbal conveyance of acceptance and presence.*

PRACTICE: Think of a time you were having difficulty holding yourself together emotionally and someone simply offered their nonjudgmental presence. The person didn't try to solve your problem or talk you out of your feelings. Instead, the person had a loving presence conveyed through their body language. Chances are, it felt very supportive. If you can't ever recall such an experience, perhaps imagine one in your mind. Imagine the kind face of a trusted friend or even a beloved pet, letting you know through their presence that you aren't alone.

Resonating is a powerful act of self-building we can give toddlers. *It's the quiet first step of attunement.* So often we ignore a toddler's behavior or try to correct the child who is in a bottom-up state. But when we realize that this is when a child's nervous system needs connection the most, we can shift our lens and warm up our response before we say a word. As Dr. Porges says, "It's not what you say, it's how you say it."

RESPOND: The first step, the emotional component, primes the child to receive the next part: figuring out what kind of deposits to make next into the child's body budget. What does your child need to stabilize their system? If your child is old enough to answer you, the best way to find out is to ask them, "What can I do to help you?" or "What do you need, sweetheart?" If the child can't answer, then ask yourself: Does my child need something basic that they might not be aware of, such as a snack or a nap? Do they need a gesture, maybe a hug? (Review the notes you took in Chapter 6 about the sensory strategies that calm your child.) Does your child need some soothing words? Now we can respond with language the child understands and in an empathic tone, *describing or symbolizing* the child's experience.

Once the child is less agitated, you can *attach simple words* to help the child understand their emotions, or the situation, in order to feel better. Ask yourself what might have thrown the child into the red. Did the child expect one thing and then get another—as Jordan expected milk and got water instead? *If so, we can talk them through the upsetting experience.* "Oh goodness, you were expecting one thing and you got another! What a surprise that is!" Or "Oh, I see this is so upsetting. You weren't expecting to or didn't want to (fill in the blank)." Or try gently mentioning what you think the trigger was: "Oh dear, your cookie dropped. That was unexpected!" Or "I see how frustrating this is, sweetheart. You really don't want to leave the party right now." The point is simply to recognize that there is a problem and put some words to it, compassionately framing the situation in neutral terms while staying calm.

The words help the child begin to use concepts to understand their experience. The purpose isn't to *solve* the situation but to help the child feel our attunement in a more detailed way.

EXAMPLE: Upon seeing what's for supper, your four-year-old begins to cry and announces, "I'm not eating any dinner. No way!" You might be tempted to say something like, "Oh, yes, you are!"

or "No dinner, no TV!" Or maybe "You should be grateful there's something to eat!" But rather than taking the words personally, try resonating with the energy behind the emotion. Go beyond the behavior to the trigger: a big feeling of dysregulation.

RESONATE: Center yourself and *look at or toward your child with a face that recognizes the disappointment or the struggle.* Resonate with your face, your body language, your simple words. Whether the child's platform is weak because of the additional stress of the day, or too little sleep the night before, or something else, we rely on resonating with the emotional message *I see you and I witness this struggle.* We want our children to feel validated for their emotions and feelings because we want them to value themselves and not feel bad for having ideas, desires, and opinions that are different than ours.

RESPOND: "I see how disappointed you are. You thought we were having something else." (Guess what you think threw the child, such as an unexpected change—for example, getting fish when they expected macaroni and cheese.) "That's hard, I can see. I wonder if you can find a way to eat some of it." Or "Maybe let's take a little break and come back to the food in a few minutes." Or "There are three things on the plate, so perhaps if you are hungry there's something you can try. Take your time." Tailor your response to what works for your child in that moment.

EXAMPLE: Your three-year-old, upset that an older child was allowed two scoops of ice cream when they got only one, loses control and begins yelling and hitting their sibling. Your instinct might be to explain, "That's because they're three years older than you," or say, "You should be happy you got ice cream at all!"

RESONATE: First, make sure that the older child is okay, assuring safety for all involved. Then provide a boundary by letting the three-year-old know that it's not okay to hit others, if that's in line with your parenting values. Next, through your emotional tone and look on your face, acknowledge the strong reaction and remember that their striking out means that their platform is shaky and they're in the red pathway, the fight-or-flight mode. You may add a verbal acknowledgment or may choose not to. Witness the distress in their body and brain; it's reflecting a subconsciously driven protective reaction that's not logical or modulated, a common situation for toddlers.

RESPOND: Offer a message affirming your child's idea or feeling. "I can see how much you wanted the same amount as your brother." Or "You were upset that you didn't get as much." Then pause and see how they respond. Still, hold the line on the ice cream, maintaining your parental authority and offering additional language for them to use in the future. *We can validate, contextualize, resonate, and set limits all at the same time.*

Fear of the Barking Dog: An Example of Resonating and Responding

One mom I worked with used the resonate-and-respond tool to help her toddler daughter overcome a fear of barking dogs she'd had ever since she was startled by a loud dog as an infant. The girl would cry whenever she heard a dog bark—a classic body-up reaction. Over a few parent meetings, we had discussed how toddlers come to master their fears, and when they visited my office, we tried it out. We opened a window, and her mom held her and said, "Let's listen." Sure enough, we soon heard a bark in the distance (I happened to know there were often dogs in the area), and the toddler shut her eyes, put one hand by her ear, and started whimpering. The mom *resonated* with a look of gentle concern on her face and then also mirrored the child's reaction by placing a hand by her own ear (nonverbal, emotional resonance).

The toddler then looked at her mom, who *responded* in a sweet and caring voice, "Mm-hmm, yes, doggy barking." The child looked at her mother, a more relaxed look on her face, taking her hand down. Then she looked out the window again. The dog barked again, and the child again put her hand to her ear and quickly looked at her mom. Mom did the same thing: gentle, soothing, *resonating* first with her *empathic emotional tone*, and then responding with, "Yes, doggy barking." Then she added words like "loud doggy" to her response, as the child waited for the next bark, covered her ears, and looked at her mom, over and over again. A few days later, the mother reported that her daughter had continued to put her hand to her ear and say "doggy" many times a day for the next several days. She was reenacting and gaining mastery over the stressful event. A few weeks later, she reported that instead of whimpering and holding her ear, the tot was now excitedly pointing dogs out when she saw them out the window, and even on walks saying "doggy," but not crying.

Simple as that exchange might seem, it was highly complex. The mom helped her daughter by first witnessing her struggle, and then framing it with words (*doggy*, *bark*, *loud*). Through her emotional attunement of *resonating* first and *responding* second, she helped transform the child's fear. When the child heard the dog bark again, she used *words* and *gestures* to regulate herself (pointing and saying *doggy*) and her actions showed that she was no longer afraid.

Discover What Works for Your Child

If your child is in a body-up reaction, take some time to discover how your child responds to different kinds of input from you. You can tailor your interactions to your child's sensory preferences. Some of the primary ways we soothe children:

PROXIMITY: It takes time and experimenting to discover the body-up strategies that work best for your child. When your toddler is in distress, do they prefer that you hold them, or stand close? Or perhaps they feel more comforted when you stand a few feet away, or even on the other side of a room or a door, offering your calm, gentle presence. You find out through trial and error.

VOCAL AND EMOTIONAL TONE: With your emotional tone and voice, you offer cues of safety personalized to your child, providing the message that "You're not alone. Mommy (or Daddy) is here." "I'm with you in this." "I'm not judging you for this big reaction." What *qualities* of your voice are most helpful when *your* child explodes? (Singsongy? Quiet? Whispering? Neutral? Sturdy?) Perhaps what helps is a word or two—or simply a sound that says to the child, "I get you, I see you, and I'm not judging you."

THE RIGHT TOUCH: What kinds of touch calm your child down when they're struggling? Examples: light touch on the arm or hand, firm or strong hugs, a hand on the shoulder or forehead, a special blanket or toy to hold, or perhaps no touch.

MOTION: What types of movement calm your child when they're upset? Does your child like to be held or rocked? Need room to move about safely on their own? Prefer to be in their own space with little movement?

VISUAL CUES AND CONTACT: What kinds of visual support help your child to calm? Does your child like to have eye contact with you? See a gentle look on your face? Or perhaps your child needs to look away or even have you look away.

YOUR WORDS AND IDEAS: Words are how we build bridges between our ideas and those of others and provide children with tools to master their concerns and fears. You can experiment to see what words resonate with your child in different circumstances by looking at the level of tension and calmness or agitation in their body, and in their feeling tone. Recognize that sometimes when we tell children how we think they are feeling, they may respond defensively. Choose the words that work best for your child.

Some other ideas to consider to help alleviate body-up behaviors:

LOOK FOR EARLY SIGNS OF FATIGUE AND THE PLATFORM GETTING SHAKY. Draw closer to the child emotionally, letting your child "borrow" some of your own regulation ahead of time. Being more engaged with your child can help lessen the child's stress load, sometimes heading off a full-blown tantrum.

MAKE YOUR PLANS FLEXIBLE. If you see your toddler starting to fuss, whine, and show signs of stress in their body, adjust your expectations and pull back on what you are asking of the child. The best time to ask the child to push their limits is when they have a nice big balance in their body budget (a good night's sleep, food in the tummy, feeling safe and secure).

GIVE YOUR CHILD CONTROL OVER SMALL THINGS. Toddlers love to make decisions and to have control in their environments. We want them to flex these muscles, so allow your toddler to help in decisions that are appropriate to your situation. Allow yours to create or play with objects or food in the kitchen alongside you, or decide the order in which to eat meals, or where to put toys "to bed" at night. Give them the chance to feel how good it is to make decisions independent from your directives.

Solutions for Top-Down Toddler Behaviors

Toddlers don't have sophisticated high-level thinking abilities yet, but day by day they gain more control over their bodies and interact with the world through their gestures and words.

We see it when they say, "Mommy, can you sit by me," and put their head on your shoulder, instead of whimpering or crying by themselves when they feel anxious. Day by day, month by month, responsive interactions help toddlers to build these top-down abilities.

Yes, *sometimes*, when their stars align, toddlers can ask politely instead of grabbing, or tell you that they would prefer not to eat broccoli right now instead of throwing it. But these abilities are *emerging* and not yet reliable.

Top-Down Abilities Emerge Little by Little

Jenna, age three, was crying and refused to go to bed. Separating from her mom at nighttime was a constant challenge. She frequently got to travel with her mother, a musician. Sharing a room with her mother, she fell asleep easily, even if they didn't share a bed. But at home, Jenna called her mom into her room every few minutes for yet another glass of water and yet another question or request. Lying in bed, she would feel overwhelmed by the need for her mother and didn't yet have many tools to calm herself, other than to ask for things—a top-down behavior.

The ability to use thoughts to calm her body down was a top-down skill that was still emerging for Jenna. *Planning actions, making decisions, comparing choices, holding back on impulses—these are skills that take a long time to fully develop.* Children don't magically acquire these abilities at a certain age; they evolve with the child's development and support from the adults around them. Over time, children gain the ability to use their own strategies, such as ways to move their bodies or use their thoughts to solve problems for themselves.

We want to provide children with the time and space to create and experiment with different options and to exercise their emerging muscles of self-regulation. We use empathic limit-setting, helping the child to feel safe and not ashamed or blamed for exploring the way their brain compels them to. We also support their growing ability to debate using their top-down thinking, without giving up our authority and leadership. We can let the child understand that it's okay to have a separate opinion and disagree with us, validating their growing independence. When we shut a child down or reject them for disagreeing with us, it communicates to them that they have to be a certain way (positive, compliant) in order to win our acceptance.

Some examples:

Little Scientist

EXAMPLE: Your two-year-old dumps the contents of your makeup bag on the floor. You're tempted to get mad, but you remember that to a toddler, *everything* is a science experiment, so you say something like, "Oh my goodness, you found Mommy's makeup! This isn't for you to play with, but how about if we go find some other stuff to play with?" You need not be negative or harsh. You can certainly set limits, but remember that exploring their environment is a natural part of being a toddler.

Controlled Acts of Misbehavior

EXAMPLE: Your four-year-old stealthily takes the teething crackers off his little brother's plate, then scoots under the table to eat them in secret. You're tempted to reprimand immediately, but you remember that this is an age to experiment with social problem-solving and understanding other people's emotions. So instead of, "Drop that cookie right now!" you might say, "I see that you took your brother's cookie but didn't ask first. Hmm, that's not how we treat each other in our family [or whatever value you are wanting to share with your child]." *The purpose is to notice and communicate your family* *value or lesson appropriate to the situation, but not to shame the child so harshly that they shut down.*

When a toddler takes his brother's cracker, it's not really a misbehavior; it's learning about how the world works. *Toddlers are only just learning to put themselves in another person's shoes.* They often need our help to stop and think about the consequences of what they did based on the reactions of the people around them. Toddlers want to explore and learn by trying out all sorts of behaviors to see how they work and what the consequences are. We calmly address these behaviors and try to promote reflection and learning moments by working with this natural part of development.

Negotiating, Bargaining, Pushing Limits

EXAMPLE: Forgetting that you let your three-year-old pull the lid from her favorite yogurt the last time you served it, you hand her the yogurt cup without the top. Upset, she insists that *she* wants to open it herself and demands that you get a new container out of the refrigerator so she can do so (an admirable act of negotiating and problem-solving). You don't want to open a new container, even to avoid a power struggle. Instead of saying, "Too bad, deal with it," you *engage her in a conversation to help her manage the disappointment.*

"I see this means a lot to you," you say. "Last time you got to open it yourself and this time you hoped you could do it again. It's okay to feel disappointed." You pause and let the child process what you have said. Then prepare to co-regulate with the emotional material and words your toddler serves back to you. If the child is still negotiating—wonderful. She's honing her debate skills! Using words means that she is expanding her knowledge and vocabulary to manage her emotions in the future, and you have a lot of material to work with. Be patient and remember that you can still empathize with your child without agreeing or giving her what she wants. *Giving in when you know doing so isn't in your child's best interest doesn't build resilience, it degrades it—because it denies the child the chance to experience and tolerate disappointment.* On the other hand, we can appreciate that humans like to feel in control, so we can understand the emotion and the intention without relinquishing our authority as parents.

Toddlers are enamored with the power of communication and self-direction. They go from being babies—dependent on others guessing what they need or how they feel—to being able to communicate through pointing and gesturing, and later in childhood or adolescence to using words to symbolize their internal world and their emotions. It's invigorating and liberating to feel like the world is in your control, but then disappointing to discover that there are still many things you can't do.

Play: A Powerful Tool to Help Build Top-Down Abilities

In Chapter 7, we discussed the importance of three words that guide our play with children: *Follow their lead.* Doing so gives us insight into the child's interests, intentions, and motivations. Play in the early toddler years is all about exploration, letting children explore their surroundings while making sure that they are safe. This loose definition of play includes dumping pots and pans out of kitchen cabinets and dropping food repeatedly from the high chair. Following a child and seeing the world from the child's perspective is an exercise in being present and observing life in a fresh new way. *Taking in a toddler's self-directed exploration gives us the benefit of being present and experiencing joy in unexpected new ways.*

Patience for Repetition

Toddlers love repetition and discovering how things work. Opening, closing, taking out, putting back in, and then repeating it! Why? All humans love to predict or expect what will happen. It makes us feel safe. So a child finds it satisfying to open and close a container *over and over* again. But there's more going on: By repeating things, a child is learning all about their physical world. They're also becoming familiar with patterns, and pattern recognition is a skill that's found in mathematical and other forms of reasoning. So try to have patience and understand that spilling and dumping and repeating aren't useless; they're all in the job description of robust toddlers using their bodies and brains to explore their environment. *And the best kind of predictability comes from parents whose emotional reactions are predictable and stable.*

Toddlers develop from infancy when they go from batting at things to mouthing things to handing you things to human interactions and the serve-and-return of more active play. They show you a ball, and you respond with a look of interest. They may hand you the ball, and you say, "Ball!" and then hand it back and wait to see what they do next. Perhaps they hand it back or drop it, waiting to see what you do next. In this early part of sensory play, in the second part of the first year, it's wonderful to prioritize the back-and-forth and see where it goes rather than imposing limits when you are playing. Remember that the toddler is a scientist, and everything is new territory to be explored. Dropping, squishing, mixing food together, and seeing the colors change—these are all lessons in the real world.

As the toddler years progress, the play goes from sensory exploration to using toys realistically: rocking a doll or rolling a car back and forth. From there, a child's play evolves to the ideas and themes that interest that particular toddler. Play becomes one of our best ways of discovering what is on the child's mind and what kinds of things they're worried, curious, or concerned about.

That's why I chose a play-based approach to help Jordan, who was having such difficulty in preschool. His parents understood that experiences in early childhood had an impact. When he had lacked sufficient co-regulation, his ability to self-regulate diminished. So the best way to help him develop control over his behaviors wasn't to give him more rules but to give him play-based opportunities to practice *modulating* or learning how to control his reactions to disappointments. Jordan needed lots of play-based and everyday co-regulation to build up his confidence in the world as a safe place—which had been absent during his infancy and early toddlerhood. Eventually, with the unwavering support (and advocacy) of his parents, Jordan found success and safety as a beloved son and respected elementary-school student.

His experience illustrates how play can help children understand and explore their world and their emotions, helping to build self-regulation. (In the next chapter, we'll examine what play reveals about the child's inner world.) On the other hand, toddlers like the one I saw at the airport have bottom-up reactions because they lack the skills to help themselves tolerate experiences such as fatigue or hunger—they still need co-regulation to do that. He was too early in his development to know that he was hungry or tired and ask his parents for a snack, or understand that he would see his Fluffy Bear again in a few hours, when their flight landed. We build these skills through our relationships: how we interact, how we play, all of the interactions that provide your child with the opportunity to use words to tolerate a wide range of discomfort, develop better self-regulation, and talk about what they feel or need instead of melting down. As children progress out of toddlerhood with a sturdy foundation that came from co-regulation, we can now discover a whole range of ways to teach children about the wonders of the feedback from their bodies to their brains—our next topic.

RESILIENCE-BUILDING TIP: Toddlers begin to develop self-regulation when parents have expectations and offer challenges and appropriate support in line with their children's emerging abilities to control their emotions and behaviors. Exploratory play is the most natural way we support a toddler's development.

9

Elementary School–Age Kids

FLEXIBILITY AND CREATING A BIG TOOL CHEST

> Emotional literacy helps your emotions work for you and not against you.
>
> —Claude Steiner

Sometimes we think we can make our children's future lives more secure by providing them with resources: toys, tutoring, tuition. These are all nice. But our most important job as parents is ensuring our children's future emotional health. To do that, we need to nurture in them an essential trait: flexibility, the ability to shift in response to life's changing demands and challenges. Flexibility is a cornerstone of resilience.

When I faced emotional challenges as a child, my parents often tried to reassure me by telling me there was nothing to be afraid of and nothing to worry about. Like many parents of their generation, they thought it best to think away or ignore negative emotions rather than see them as valuable signals that were worth paying attention to. When my own children were young, I took another approach: I tried my best to co-regulate with them. I acknowledged that a loud noise was scary; I offered a compassionate ear if one of them was feeling rejected by a peer. As a result, now that they're adults, my kids are far more flexible and resilient than I was at their ages.

When our children are young, we help them by making deposits into their body budgets through co-regulation, comforting them as infants, and listening and responding as they grow. As they get older, a new priority emerges: helping them gain the ability to manage their *own* body budgets. In this chapter, we'll focus on two ways to help our children become flexible problem solvers through two main channels: the magic of play and how we talk to them about emotions.

School-age children are equipped with everything they learned through toddlerhood, but they are still developing the abilities they will need to manage life in later years. After toddlerhood, children encounter new demands as they navigate the world of school and peers, and discover more and more about themselves. It's a process, one that comes with struggles and feelings that can trigger behaviors that parents often find confusing or troubling.

In this chapter, we'll discuss how to nurture *top-down skills*—thinking, problem-solving, flexing to cope with the unpredictable—on top of the solid base of trust and safety. We can help children learn to use their minds to help their bodies support them as they discover how to advocate for themselves. The goal isn't to teach our children to memorize the "correct" responses and answers to every problem they'll encounter. *It's to teach them how to help themselves, and develop mental flexibility in order to use their brain to calm and direct their thinking—the focus of this chapter.*

We'll see how play can be a powerful tool to support a child's overall development. And we'll learn how to talk to children about their emotional health in ways that support their awareness and understanding of their bodily sensations. In short, we'll explore a primary source of resilience and the cornerstone of my field of psychology: the ability to use your mind to change how you think and feel.

SIBLING RIVALRY

Alan and Camilla were having difficulty managing their children's sibling rivalry. Mira, seven, and Leo, four, constantly bickered and argued over toys and other possessions. The parents described Mira as a sensitive child who loved school and did well, routinely helping out her teacher and fellow first-graders. She thrived on predictability, rules, and structure.

It was when Mira returned home each day, though, that "all hell breaks loose," as her dad explained. After excitedly greeting her, Leo would grab papers from her backpack, running away with or tearing them, prompting Mira to scream and cry over the worksheet or art project he'd wrecked. She would appeal to her dad, who would chase after Leo, but the toddler approached the whole scenario as a game. Finally Mira would lose control, pushing her brother and taking back her things. Predictably, Leo would burst into tears. Over time, Mira found a solution: She turned bossy, watching her brother's every move in order to control the situation. Tensions ran high in the family, with the siblings locked into squabbles, rendering their home a constant battle zone.

I reassured their parents that sibling rivalry is an expected, albeit challenging, part of life, and even has its benefits. Siblings provide each other with ongoing opportunities to learn how to solve problems cooperatively. Still, the home's emotional tone was negatively charged, and I sensed that both children were immature in their problem-solving. So I suggested I evaluate each sibling individually to see if I could offer suggestions to expand their challenge zones and ability to work things out on their own, bring more balance, and introduce some joyful sibling play.

I focused first on discovering the right approach for the family and for each child. As it turned out, Leo and Mira both needed the kind of help for which play is perfectly suited—building up their tolerance for frustration and ability to navigate their tussles using their words. Eventually, Mira also needed a more direct approach: discussing her top-down understanding of her bottom-up feelings. In other words, she was learning to use problem-solving skills—including active self-regulation and language—to manage her frustrations, figure out solutions, and exercise her social and emotional muscles.

I introduced the parents to my approach, and they felt some hope as they came to understand that they were dealing with normative behavior that they could understand better. I suggested we start with the best possible (and most underutilized) building block of children's problem-solving skills: play.

SYMBOLIC OR PRETEND PLAY: A POWERHOUSE NUTRIENT THAT SUPPORTS ALL ASPECTS OF DEVELOPMENT

In decades of playing therapeutically with children and their parents, I've learned important lessons: When we follow a child's lead in play, it shows us what's on the child's mind, what the child is worried about, and where the child needs extra support. Play is also an organic and effective way to nurture a child's emerging capacity for executive thinking. Play is the way children build skills from the ground up. *If you've tried to talk to your struggling child about difficult subjects and didn't get far, then playing with your child might be the best place to start.*

Play is recognized as such a powerhouse of development that the American Academy of Pediatrics recently recommended that pediatricians prescribe play at well-child visits. I've witnessed this magic over and over again. Play fosters children's ability to talk about feelings rather than blow up and to solve problems by practicing safely with the themes or conflicts they're facing. Parent-mediated play, as it's known, helps children cope with sibling rivalry, anxiety, and more serious challenges, such as stress and trauma, serious medical conditions, and losing a loved one.

Play also supports the executive skills that we want our children to develop. As the clinical report to pediatricians put it, "research demonstrates that developmentally appropriate play with parents and peers is a singular opportunity to promote the social-emotional, cognitive, language, and self-regulation skills that build executive function and a pro-social brain." That's powerful.

A pro-social brain is one that helps a person be cooperative, see others' points of view, and solve problems—exactly the abilities that would help children like Mira and Leo navigate and manage the tensions between them. A pro-social brain is also what enables children to engage on the playground and in the classroom, participate in group projects and academic interactions, join team sports, and manage the complexities of social life in high school and college.

Another study found that the benefits of active play extend to school and other key parts of development. Research showed that high-quality school recess contributes significantly to children's executive functioning (focusing, making plans, and solving problems efficiently) as well as self-control, positive classroom behavior, and resilience. Children who had more time to play at recess were better able to overcome adversity, recover from mistakes, and cope with change.

Our culture has a bias toward classroom instruction over play. *But I have long observed that children who play more are happier, less stressed, and more creative in how they solve problems.* With those benefits in mind, let's examine some ways to bring this beneficial type of play into your family's life.

Play Basics

Make Time for Your Child to Play

Since research clearly shows that play benefits children, it's important to build it into your child's life. If your child is in preschool, make sure the child's program includes plenty of *unstructured* playtime. What helps young children learn to solve problems and self-regulate isn't structured academics or group learning but playing. If, like many traditional schools, your child's school doesn't prioritize play, then try to offer opportunities to play with friends or extended family regularly. Then sit back, make sure the kids are safe, and try not to interrupt the play (unless someone is at risk of injury or intervention is clearly necessary). Children practice their problem-solving in free, uninterrupted play, and they need time to experiment and explore what works. They develop creativity and gain confidence in their abilities starting early. Watching a child play on her own recently, I couldn't tell what was driving her ideas until I saw that she had organized some toys on a table in a way that resembled a dinner-table setting, while beaming and saying to herself, "I did it!" Independent play and exploration build resilience. Let them be.

But there's another kind of interactive play in which *you* can engage with your child. Earlier, we discussed exploratory play in infancy and the value of peek-a-boo, one of the earliest ways adults play with babies. We've also described how toddlers use their senses to explore their world like scientists. What emerges soon after is *pretend play*, also known as *symbolic play*, when children use toys or themselves as characters or participants in dramas they create. This mode of play offers a unique and wonderful window into your child's inner world.

Parent-Mediated Play

Pretend play provides natural opportunities for children to explore, practice, and work through what they are managing emotionally. As we follow a child's lead, play reveals the issues and concerns on a child's mind, providing clues to the areas in which they need extra support or work.

I was fortunate to learn about *parent-mediated* play (in which parents play with their children) from Dr. Serena Wieder, one of the world's leading experts on the therapeutic use of play with parents. She helped me to see the value of trusting in the play process, always remembering simply to follow a child's lead and see where things go.

While play with peers has many benefits, including stimulating cognitive development in preschoolers, and solitary play helps babies and toddlers discover their bodies by exploring their world, parent-mediated play offers opportunities to get to know your child in an intimate way through the child's most natural first language. This kind of play offers a nonthreatening way to help a child explore a range of natural and expected emotions, such as anger, jealousy, empathy, positive feelings, and competition.

It's worth mentioning that this type of play has its basis and research in more industrialized countries. As with everything in this book, honoring your parental instincts, culture, and values should take precedence. If the play I describe here doesn't fit for your family—that's perfectly fine. There's a wide range of what humans find playful, and the benefits arise from our individual differences and variations in cultural values, norms, and beliefs.

Don't Stress—This Can Take Just Minutes a Day

All parents have busy lives, but this kind of play doesn't require a significant time investment. Even *five minutes a day* of uninterrupted play with your child can be meaningful, and the benefits accrue over time. Besides, if you indulge your inner child, you, too, will reap play's joyful, anti-stress, and anti-inflammatory benefits. The key for those extra health benefits is that the experience is enjoyable for you and your child. Of course, play looks and feels different depending on your child's age and developmental stage. With a preadolescent, it might mean laughing and chatting while you stroll through the shopping mall or share a treat at a coffee shop. But it all starts with pretend play.

One benefit of this kind of play is that it's simply easier to discover what's on a child's mind by playing with them than asking them directly about tender subjects. When I met Mira, for example, her parents prompted her to tell me about an incident the night before, when she had shoved her brother. She flashed a sheepish, embarrassed look, and simply said, "I forgot." Clearly, she felt put on the spot. While I knew the parents meant well, they had pushed her shame button—not the way to yield the most useful information!

Instead of point-blank questions, it's better to aim our attention below the tip of the iceberg, to explore and understand a child's motivations and triggers first through playful engagement and trust building. Later, once we've built that connection, we can have more direct conversations.

As it happened, Mira was early in her process of mastering using words to describe her actions and motivations, so I realized quickly that we should start with playing, which would give her the opportunity to exercise emotional and problem-solving muscles so she could eventually use words to discuss feelings and ideas.

The Basics of Parent-Mediated Play

What you'll need for parent-mediated play:

- **YOURSELF:** *If you can unplug for at least five minutes (or longer, if possible). If your child is struggling with challenges, a twenty-minute block is generally an adequate amount of time to allow a theme to emerge. If you can relax, ignore the to-do list in your head, and bring your curiosity, reflection, and some playfulness,* *you'll open the way for your child to share their world with you in whatever way they choose.*

- **TOYS:** *If it's possible and within your resources, have a few basic toys on hand. They don't need to be fancy. And toys aren't an absolute requirement—other people are the best toy a child can have! Children have imaginations and if we allow ourselves to shift our roles, we can become a "toy"—a kitty or a lion, a princess or a king. One creative family I worked with made simple dolls out of used socks, with buttons for the eyes and thread-embroidered mouths. The generous kids made me a simple "family" of the sock dolls, which I treasure and which many children visiting my office have used for various roles over the years. There's no need for expensive or mechanical toys for the play I'm about to describe, just objects that stimulate imagination and inspire* role reversals *and* reenactment, *according to each child's interests. And while board games and puzzles can be enjoyable activities (and I recommend them for family activities if you enjoy them), they don't stimulate a child's emotional themes like the more basic toys.*

- **TOY ANIMALS (STUFFED OR OTHERWISE):** *It's useful to have on hand some friendly and nonthreatening animal toys such as teddy bears or puppy dogs, and some less gentle—perhaps more imposing dogs, a bear, a tiger, or even dinosaurs. The animals can be figurines, plush dolls, plastic, wooden—anything that might appeal to your child's interests and inspire or draw out a range of emotions. You don't want only fluffy kittens or sweet puppies, because a toy lion or bobcat can help inspire a child to play with themes of safety and threat—feelings that all humans face. Some children might shy from getting involved in the play if the human figurines are too realistic looking—it might just make the play feel too close to reality. The point is that animals are often easier for* *kids to pretend with because they are a bit removed from their personal experience, but kids can still project emotions onto them.*

- **INANIMATE OBJECTS, SUCH AS CARS, TRUCKS, TRAINS:** *Playing with these can also elicit themes and provide pathways for children to show you what's on their mind. Cars and trucks can have "feelings," compete with each other, go on adventures, go to school or Grandma's house, or crash into each other. In other words, children can project human qualities onto them. And of course many kids just like playing with them.*
- **FIGURINES OF PEOPLE OR SUPERHEROES,** *such as a boy and girl and more androgynous dolls, baby dolls, or other humanlike figurines or dolls. If possible, have figurines that look like a mommy, daddy, baby, and siblings. A simple dollhouse is helpful but not necessary. You can also use simple props from the kitchen—paper plates, plastic spoons, small bowls, or anything else—as household props that inspire children to "become" a mommy or daddy or baby or sibling in the play.*

Follow the Child's Lead

After making yourself available to your child with a few simple toys on hand, then simply sit and wait, bearing in mind the magic three words we discussed in Chapter 7: *Follow their lead.* Watch what your child does with the time spent with you. They may not pick up toys at all and simply start to talk or ask questions. That's great! Listen to what's on their mind. When you make yourself available to a child without an agenda, you can learn so much from what comes up spontaneously. Sooner or later, though, your child will likely start to play with you, and that's worth waiting for.

If your child picks up a toy, don't try to direct the play, just follow your child's lead. Most important, resist the temptation to ask questions—especially questions that you already know the answers to (e.g., Do you know the name of that animal? What color is the car? How many puppies do you see?). *In play, our goal isn't to teach the child concrete concepts but to let the child explore their pretend world, that is, to experience it symbolically.* It's what kids naturally do. So wait to see what your child does with the object they pick up, and then interact if they engage you in the play in any way.

Perhaps they'll pick up a cat and put it in their lap, saying, "Here's the mommy," and then hand you a "baby" kitty. Perhaps they'll pick up a car and give you one as well. Bingo! The play has begun. That's your cue to get into character. Perhaps your kitty lets out a meow, or your car says "vroom," letting your child know you have joined them in their pretend world. Now the magic starts!

Wait to see what your child does next. If your child does or says something with their character, you respond. Maybe their kitty says "Bye-bye." They've brought in a play *theme*. A kitty or car, or any other character, saying goodbye evokes a theme related to separation. If this theme appears over and over, then your child may be working emotionally on separating from you or others. Perhaps you use your best kitty voice to meow "Bye-bye." Stay in the character so that your child stays in character, working whatever angles they need through spontaneous play. Just wait to see what their character does and then do something in response to build on the theme.

Expand on the Play Themes in Small Doses

As you continue to follow your child's lead, you can expand and deepen the play to discover even more. Add a simple question or reflection. Remember, the answer shouldn't be something you already know, but something the child can create. Instead of asking a question you know the answer to, such as "What color is the kitty?" speak in a kitty voice and say, "Hey, little kitty, where are you going?," *expanding* the play. You're asking a question in order to get additional information about what's in the child's mind. Let's say your child answers, "To kitty school." You've hit pay dirt again! Now you're privy to something else on the child's mind: school.

From there, continue to expand. Simply be patient and see what happens next: Does the kitty go to school? Does the kitty refuse to go? Is the kitty happy or sad at school? What happens at school? Simply follow the kitty's lead. Perhaps the child will hand you a larger toy cat and ask you to be the teacher. Or they'll hand you a smaller cat and they'll take the teacher role. These are more opportunities for windows into the child's internal world. Go with it. What do the teacher and the student kitties say or do? Of course, there's no right or wrong answer. It's the journey of exploration that's valuable. It's important to stay in character and not break the spell of play. Don't suddenly ask, "Darling, are you afraid of school?" Instead, stay in the pretend play, going with the child's flow. There will be time for asking specific questions later—maybe during a walk, or when you're relaxing together. *But while we're playing, we sanctify the pretend without asking our child to move outside the play.*

As you play together, follow the child and *expand* only as much as necessary to keep the serve-and-returns going. The child will work on the themes that they need to for their own development. That's what makes play so powerful. Children play out what they're working on—consciously or subconsciously.

To be clear, I'm not suggesting that parents become play therapists with their children. Your parents, however loving, probably didn't engage in this kind of play with you. I know mine didn't. If pretend play doesn't feel right, or feels stilted or uncomfortable, don't feel compelled to do it. You may find that it becomes more enjoyable over time. But if it's just not you, skip it.

If you do try it, though, you may find that play offers a powerful way to connect with your child with safety and joy—and a chance to have fun together while getting a window into the child's interests and concerns. Don't feel you have to fix anything about the child during play, just witness the child and play the various characters the child assigns you. *This is part of the magic of play: Children can experiment with concepts, ideas, and emotions outside of "real" life, but in a simulation of their own making. The power of that simulation can't be underestimated.*

Play can be a sort of reenactment that helps a child work through something they are going through by either consciously or subconsciously rehearsing new solutions to problems, or by simply rehearsing something in order to have it become more familiar—as when a child reenacts a doctor giving shots (a common theme). It's a healing process: The scary trigger becomes less scary through the powerful reenactment.

Don't Be Afraid of Negative Themes

It's natural to encounter a wide range of negative and positive emotional themes in play. This is actually a good sign; it indicates that a child is *actively working through the range of negative emotions* that aren't as easy to work on in real life. As a parent, you may be tempted to use this opportunity to *teach* your child, especially if hitting or other negative behaviors arise. But I encourage you to resist that urge. A child's aggressiveness in this kind of play won't make the child more aggressive in real life. So while you're playing, try to avoid sidebar teaching moments ("You know, it wasn't very nice of the kitty to hit the bear!"). Instead, observe *all* of the emotions your child's characters have in play and consider them a reflection of a healthy range of emotional expression.

Ideally, play will bring out a range of themes—from the positive emotions of nurturing and empathy to more negative themes, such as rivalry, jealousy, anger, and sadness. We all experience these emotions as a part of our humanity. In "real life"—at school, in their communities—children are often made to feel shame for expressing negative emotions and rewarded for positive ones. In play, though, we treat all emotions equally. It's a way to convey to the child that it's expected to feel mad, jealous, sad, or happy. Of course, it's up to you to model for your child and instill values that serve as guides as your child learns how to *act* on these emotions, outside of play. We want our children to treat themselves with equanimity, to understand that all of their emotions are valid, and not deem certain emotions as "good" or "bad."

To show you the potential benefits of this kind of play, let me share a few examples of interactions I've witnessed.

- *One child with a chronic medical condition spontaneously played hospital with his stuffed animals, pretending to transport some in an ambulance. He gave the animals he assigned to his father different roles: doctor, ambulance driver, patient. Reenacting scenarios he had experienced and alternating between "becoming" the patient and the healer gave him a powerful outlet to manage the stress of his medical reality.*

- *A child whose parents were in the midst of splitting up created separate play "homes" all over my playroom, with various characters making their own houses, complete with pillows and doll furniture. Clearly, she was setting up bedrooms and kitchens in a rehearsal for the impending move out of her family's shared home.*

- *A child who was struggling with being bullied at school had dinosaurs fighting with each other, with superheroes overcoming bad guys. When the child became a superhero in play, he could imagine and feel the power of what it means to have strength in real-life situations. Imagining the feeling of being strong offered him a way to counteract the anxiety he's experiencing at school and to simulate and practice solving problems.*

- *A child who had survived an automobile accident reenacted cars crashing into each other over and over again as well as waking up with nightmares. She also scoured my playroom for cars that resembled the vehicle that she had been riding in when the car accident occurred. Her mother actually felt triggered by the play, stepping aside and asking me to play in her place, which was a self-compassionate decision. After several months of acting out car crashes, the child shifted to other topics. She had processed some of her trauma through play, and it no longer occupied such a significant place in her mind. That progress was also reflected at home in her improved sleep and end of the nightmares.*

THE BENEFITS OF PLAY

Through play, children act, reenact, and experience many different roles, then reverse their roles, so they can experiment with all sides of an issue in order to better understand themselves or others. Play is a pathway toward creative expression and working through issues. It provides a space in which your child can express a range of emotions, where it's okay for the pretend characters to experience anger, sadness, jealousy, joy, nurturing—the whole gamut of human experience. *We call play a neural, or brain, exercise because it's a chance for children to process different sensations, feelings, and ideas under conditions of safety.*

As children play, they develop their future ability to talk about feelings and ideas related to stressful and joyous events and problems in real life. We want to support and strengthen this ability through play, because it is the natural language of childhood.

An important disclaimer: If you and your child aren't enjoying yourselves, or if you feel that play is depleting your body's budget, or your child's, then it may not be useful. That's okay! If you feel in over your head, or if the themes that arise are triggering for your child or if you have serious concerns about themes that emerge, then it's useful to contact a child therapist who is familiar with developmental play. Keep in mind: Play should *not* add to your stress load or strain your relationship with your child.

Play should be fun for parent and child alike. Don't analyze, fix things, or teach. Sure, you can learn from playing with your child, but the real benefit is to have fun together, give your child your full attention, and have joyful moments of presence that are based on a free agenda. That's truly a gift to your child, especially in our overscheduled and academically focused world.

Playing with Leo and Mira

I learn a great deal about families by watching how they play together. When I watched Mira and Leo's family play, Leo was hesitant at first, staying close to his mom as Mira and her dad began exploring the room. Before long, Leo, too, was looking around and trying out a toddler-size slide and then exploring my large playhouse, a structure that I've watched various children use as a café, a doctor's office, a veterinarian's office, a school, a day care center, and a market, among other things. *This is the beauty of play: Children make the toys or objects what they need them to be.* So a playhouse or even a big cardboard box can inspire countless scenarios.

Mira decided it was a McDonald's and enlisted Leo as the cook, a role he gladly accepted, clearly delighted that his big sister had given him a task. She asked her parents, Alan and Camilla, to be the customers, and they played along as hungry patrons, even making special orders. Seeing how easily they all slipped into character, I immediately saw the potential of play to help their family, giving the children ways to explore different parts of themselves, their emotions, concerns, and conflicts.

The cooperation and role-playing continued until the inevitable rupture happened. This often occurs toward the end of a session, when everyone is feeling comfortable and letting their guards down. Tired of the McDonald's scenario, Mira had wandered to a corner of the room, where she found a bucket of toy frogs and settled into arranging them in a circle. She hadn't gotten far when Leo ran over and grabbed some of the frogs. "He messed up the frog circle! Leo, no!" she yelled, crying and looking to her parents for help. Both tried intervening, but Leo was off to another part of the office, frogs in tow.

Over the next several sessions, the sibling conflicts arose again and again, with Mira and Leo playing cooperatively, then Leo doing something unpredictable and his sister erupting into anger and tears and asking their parents for help or angrily grabbing the toys back. I advised the parents to continue to support the children as they worked out their problems on their own, and also had them return with one child at a time to give us a chance to focus on the children's individual issues.

Leo: Using Play to Develop Better Self-Control (and Deal with Sibling Rivalry)

I hoped to focus on each child's challenge zone, helping them to increase their ability to negotiate and solve problems in better ways than fighting or bickering. My goal wasn't to eliminate the sibling conflict but rather to help both children increase their tolerance for frustration and their ability to use language to communicate with each other more effectively. Since Mira and Leo were at such different development levels, I started with a *bottom-up play approach* for Leo and a more *top-down problem-solving approach* for Mira.

Leo: Using Play to Bolster Sibling Play and Cooperation

Without Mira present, Leo was far more cooperative. That made sense: His parents played well and didn't represent the kind of stress or competition he associated with his sister. To move more toward his healthy challenge zone, I coached his parents in therapeutic play. In this more direct form of play, I encouraged them to expand on his themes, playfully introducing opportunities for him to experience a wider range of emotions and in turn build up tolerance.

Since Leo loved playing with the bucket of frogs, I encouraged his mom to take on the role of a sibling or peer, expanding on Leo's play while adding a mild stressor. If his character was the baby frog, she became a sister; if he chose to be the daddy frog, she was the frog's child. I coached her to playfully challenge Leo in order to help him work the muscle of self-control. In the play, Leo could reverse roles.

If Leo was a little frog, then Mommy took another frog, became a sibling, and did something to provoke Leo's frog character in a playful way—for example, stealing some pretend food and running away with it. This moved Leo into flexing his muscles of self-regulation in a safe and fun way to manage the situation. If he needed assistance, his mom and I acted as coaches, helping him to reverse roles.

At the first couple of sessions, each time his mom's frog took something from Leo, he simply grabbed it back or even attacked his mom's frog. But at the third session, Mommy, in the frog's character, gently prompted him to try another response. "Do you want to tell me something, little froggy brother?" she asked, looking at the frog. "Yes," Leo said, in character. "Don't take my frog food!" Success! He had created a solution using *words* rather than simply physical actions. That's a big step for toddlers. Leo's frog was becoming his own advocate and talking to the frog who had violated the rule. In essence, Leo symbolically became his sister, and he was able to rehearse, through pretend play, the problems going on at home, and then flex the muscles of control and self-control in play. Play allows children to feel in control and to feel that power in their bodies; no wonder they love to play!

This approach proved far more effective than what his parents had done when he had grabbed his sister's toys: reprimand him or put him in a time-out corner to think about his behavior. *Play was giving him the chance to reverse roles, try different approaches, and advocate for himself.* Playing gave him the opportunity to experience and grow his frustration tolerance and social problem-solving skills. Play was a natural and developmental way to work the muscles of using his language and words and improve emotional control because it allowed him to practice by experimenting in different scenarios with different levels of calmness or agitation in his body in a fun and nonthreatening way.

Mira: Developing the Top-Down Understanding and Expression of Bottom-Up Feelings

As for his sister, Mira's play was already strong. While she played with her parents and me, her interactions were complex, nuanced, and highly developed. She could role-reverse, becoming a mom or a teacher, a child or a baby. It came as no surprise that her characters were flexible in play. She loved controlling situations and loved playing the part of a teacher or parent, gently teaching the children in those roles and demonstrating the ability to have self-control when she felt safe and wasn't threatened.

Even so, while she could play with her parents, she had difficulty using her problem-solving skills with her little brother. This told me that she was ready to learn about and practice *how to use a top-down understanding of body-up feelings* when the going got tough, such as when Leo attacked her or grabbed her possessions. Our work turned to helping her develop the powerful tool we all have to meet challenges and solve internal conflicts: using our thoughts to shift our emotions, attitudes, and behaviors. We want our children to think about what they do before they do it, and to have reasons, consistent with how we have raised them, guiding those decisions. *We want to nurture their top-down thinking, by helping them make sense of their body's signals and appreciate all they can learn from and about their emotions.*

BUILDING A NICE BIG TOOL CHEST

When children can think about different ways to solve a problem, they'll be able to create a larger number of solutions to problems they face in life. We want our kids to have many tools in their tool chest, not just one or two. And the way we help children to build up their tool chest is to help them befriend their nervous systems.

HOW DO YOU FEEL?

In Chapter 6, I described interoceptive awareness, the feeling of sensations inside the body. We have learned that "an emotion is your brain's creation of what your *bodily sensations mean*, in relation to what's going on around you in the world." Additionally, researchers suggest that linking the awareness of these internal bodily sensations leads to improved emotional regulation. If you and your child are more aware of your bodily sensations, you will have a head start in supporting your emotional and mental health. The question "How do you feel?" suddenly has a new and more powerfully nuanced meaning.

We can support children's self-regulation by helping them tune in to their body's sensations and make sense of them with self-compassion. And we now have a whole new way of supporting our children's mental and physical health, as well as our own, bringing together the body and brain in our parenting. This concept is redefining how we understand our emotions and actions.

I was fortunate to learn from mentors who had this approach long before it was a theory or was popular in neuroscience. As I've mentioned, my cross-disciplinary training was grounded in using nurturing relationships to understand how a child perceives their world through their sensory and motor systems. What I discovered in clinical practice is what researchers are finding today as they study interoception and its role in emotional thinking.

AWARENESS OF BODILY SENSATIONS AND MATCHING FEELING STATES AND EMOTIONAL WORDS

Slowing Down

The first way to help children develop awareness of bodily sensations is simply to slow down and be mindful together. This comes more easily to young children than to adults. When you have a few minutes and are feeling in the "green," sit down with your child, and follow their lead. Children have a way of marveling at things we take for granted. Nature requires being present and aware. If you live near trees, leaves, flowers, or grass, take a walk with your child, or simply sit and see what the child notices. If you don't have nature around, it's amazing what you can find growing out of cracks in the sidewalk or just by examining cloud formations.

One child I know loves to sit on the curb with me and watch the ants scurry about. One morning, she started pointing them out, and we sat together, transfixed and present in the moment. This kind of mindful presence establishes a foundation for interoceptive awareness. Take a few minutes and see what your child notices without your interference. Just be. Children experience awe in the natural world. All we need to do is provide opportunities to be in the present moment—and in the process become more mindful ourselves.

Modeling

Another way you can help your child develop awareness of bodily sensations is by *modeling* it. When you talk about your experience in the moment, as life happens, you provide a powerful example to your child. Make self-observation part of your family's conversations. When you notice a sensation, if you can, *link it to a mood or feeling state, or even an emotion.* That helps your child see you connecting an experience or physical sensation to a word, linking your body and your mind. The goal isn't to burden our children with our fears or concerns, but to show them that our sensations and feelings are our allies in managing our responses to this ever-changing world. If you aren't used to doing this, then I've just given you a head start on taking better care of your own emotional health, so *both* you and your child benefit!

Examples

- *You're driving with children in the car and a fire engine with a loud siren comes close but passes you. First link the sensation to your physical reaction: "That loud siren caught my attention!" Then ask the children, "What did you notice?" or "How did that make you feel?" The children may follow your lead and name a sensation, an emotion, or both. If they don't, that's fine, too. Next, if it works for the situation, you can link the physical sensation you felt to an emotion* **you successfully regulated**: *"I felt a little* scared, *but now that it's passed us, I feel better." You've just modeled the sensation-to-feeling-to-emotion process for them.*

- *You burn a casserole just before your guests arrive for dinner. You name the event that is causing you to notice your sensations: "Uh-oh, I smell something burning." Then, if you're feeling sturdy and can follow up with a resilience-building moment that models flexibility and in-the-moment thinking, say to your child, "I'm going to be really creative and make something else. Anyone want to help me look through the refrigerator to see what other food we have?"*
- *While you're trying to park at the supermarket, another car sneaks in front of you and nabs the space you have been patiently waiting for. "Wow, I was waiting patiently. Now my heart's beating a bit fast (or another accurate description of how you are feeling) and I'm starting to feel angry about it," you say, linking the sensation to the emotion. And then model* self-regulation *by saying, "Things like this happen. Before we go into the market, I'm going to take a few breaths and calm my body."*

When you model an awareness of your sensations in your body and the acceptance of your own positive and negative emotions, you not only increase your own emotional intelligence, you provide your child with a powerful guide to doing so themselves. Children want to be like their parents, and when your child sees you narrating your experience with a sense of self-acceptance, it also lets them see you embracing flexibility, modeling resilience at the same time.

In addition to modeling, you can begin to ask gentle questions that help your child link their sensations to their emotions. For example, if the child describes feeling jittery or "butterflies in the tummy" before a class presentation, you can tell them that those feelings are a sign that their body is helping them get ready to do the presentation. We can help a child *recast sensations* of the nervous system into signs that the body is doing its job to help the child prepare to accomplish something. They now have a head start to emotional regulation!

If you see your child beginning to have a shift in their platform, you can help them tune into their body by asking questions such as, "Sweetie, I wonder if you are feeling anything inside your body?" "Is your body giving you any feelings from your tummy, your heart, head, or anywhere else inside?" To get to the heart of emotions, ask a child if there's anywhere in their body that's "talking" to them. If the child is able to notice the sensation, it's useful to discover the valence—that is, if it feels pleasant or unpleasant. From there, you can discover if the child's sensations can be further classified into emotional words, such as *worried*, *anxious*, *scared*, or *sad*.

TOP-DOWN UNDERSTANDING AND MANAGING BOTTOM-UP FEELINGS: HELPING CHILDREN FURTHER EXPLORE THEIR NERVOUS SYSTEMS

When children can observe the state of their own platform, they gain a sense of self-control and self-regulation. Again, stopping and observing ourselves and our children is the key to co-regulating with them as parents. We want them to learn how to do this for themselves. *The goal is to help children learn to recognize, with curiosity and without judgment, when they have stepped out of a regulated state and how they can return to self-regulation by using or adding more tools to their toolbox.* Too often, children feel blamed because our culture generally frowns upon negative emotions and what are deemed as disruptive or "attention-seeking" behaviors. Teachers reward children who are positive and compliant with happy-face stickers, while tracking or sending parents notes about those who appear noncompliant. *Our society has a bias against negative emotions and the body movements that are often associated with a depleted body budget and the resultant agitated, checked-out, or otherwise misunderstood states.* Too many teachers and others lack an integrated understanding of the brain-body connection and the protective role of "negative" behaviors in keeping us safe when our safety-detection system gets triggered by a perceived threat.

Instead of *telling* a child how they're feeling, *ask* the child how they're feeling. If the child can't answer, then it's fine to reflect on what the child *might* be feeling. Simply notice, with empathy and with emotions that match the intensity of the situation, that something is happening, and then, in a nonjudgmental and open way, welcome the range of positive and negative feelings or emotions. In this way, you can help your child to have self-compassion as they recognize and own their range of feelings and emotions.

If you ask a child how they're feeling and they can reply with a descriptive emotion word, that's great. Once a child has one word to describe a feeling state generally, they can develop more to increase the detail and their emotional vocabulary. For example, a child can move from saying, "I feel bad" to a more precise and nuanced, "I feel scared (or sad or angry)." (We discussed this in Chapter 5 as "emotional granularity.") That opens the door to countless conversations about what's on a child's mind, what they are facing, and what problems they are trying to solve.

But for children who can't yet match a feeling with a word, there are ways to help. One option is to offer encouragement and a neutral, simple example from your own experience. If you think your child is jealous, for example, say, "I wonder what that's like for you. I remember when my brother used to take my toys and that made me feel jealous." *Sometimes, joining with a child and normalizing what they are feeling helps the child to resonate and open up.* If your child seems to need help, instead of saying, "I think you're mad," reflect in a gentler way: "I wonder how that feels for you. Maybe you're sad (or mad or frustrated) about that (situation)?"

Mira's parents had read that it's important to label emotions, but they found that when they pointed out to Mira that she was mad or upset, she reacted defensively. For example, if her mother said, "Oh, yes, your brother makes you mad," it would make Mira even angrier. "I'm not mad!" Mira would say. While it seems like a good idea to help children label their feelings, it can easily backfire and make a child feel *more* agitated if they haven't yet connected basic body sensations to emotionally laden words, or if they associate negative emotions with shame. Western cultures in particular connect certain emotions with negative connotations. It helps to have children come to their own conclusions about their bodies, emotions, and how they are feeling. Let's explore how to do that.

FINDING THEIR OWN SOLUTIONS

We want our children to stay socially engaged, rather than moving into a fight-or-flight mode. To that end, it's important to help children develop a language for their own platform and solutions to help the child understand their own nervous system. How do we do that?

First, I *don't recommend* using the concept of the color pathways (Chapter 3) to teach children about their own nervous systems. While these concepts are helpful to adults, one reason I don't use them with children is that educators so often use colors—in behavior charts, for example—to track behavior or try to teach children about controlling their behavior. In some of these systems, children (or their teachers) move their dots on charts to the color matching their behavior (e.g., green=good, yellow=not so good, red=bad). Unfortunately, these color charts tend to *increase* children's stress because they represent a visual threat, with children worrying about the embarrassment of having their colors "degraded" in front of their classmates. The color coding of the autonomic pathways I described in Chapter 3 have nothing to do with behavior charts. *We want children to view all feelings and emotions with equanimity, not with shame, since they are all adaptive.*

To personalize the experience and help children develop an appreciation for all of the ways our bodies show up when we are facing life's challenges, I suggest a different approach. We start by helping children develop a healthy appreciation of their own nervous system and how it protects them. We then help them organize solutions based on their own self-assessment, that is, what works for them. But we don't use fancy words. We let children choose their own words.

Consider the following script as a *general guide* to a conversation introducing the nervous system to your child. Of course, you can tailor it to your child's developmental, language, and comprehension level. Have some paper and crayons or markers on hand so that your child has the option to draw or write something about what they are experiencing, reinforcing the learning aspect. And adjust to use language that is familiar to your child.

HELPING CHILDREN APPRECIATE THEIR BODY'S SIGNALS

SCRIPT: The feelings and moods that come from inside our body are our body's way of protecting us and helping us stay healthy and balanced. (Then insert two recent examples of stressful and calming situations from the child's life or your own life, e.g., "Remember how my voice was so loud when we were late to your soccer practice last week? That wasn't your fault. I was stressed out and my voice showed it. That was my signal to slow down." Or "Remember when we baked cookies and we watched a movie together the other day? You told me you were feeling so happy inside.")

Our bodies can feel different ways and they are all useful and important. Let's think about three main ways our bodies and minds can feel. Sometimes our bodies feel calm, and when we do, we also feel happy, cozy, and safe. When we feel this way, we often want to play and do fun stuff with others. Can you give an example of a time when you felt this way? What were you doing? Can you think of a word (or words) that describes your body and mind when you feel calm, cozy, and safe? (Allow plenty of time, and then ask your child to share their special word, if they want to write the word and/or draw a picture of it.)

Now let's talk about another way humans can feel. Sometimes we feel wiggly, mad, scared, angry, or like we want to run or move—fast. When we feel this way, we might do unexpected things we later feel bad about, such as hitting or shoving, or saying something mean. We might say or do something that surprises us. Can you think of a word that describes your body and mind when you feel wiggly, angry, or like you want to get away from something or someone? (Again, offer the child the opportunity to write the word and/or draw a picture of it.)

There's another way humans can feel. Sometimes we feel sad, lonely, or slowed down. This is when our body doesn't want to move very much, and we're not interested in doing things with our friends and family, not even fun things. Sometimes we can even feel "frozen," like our body can't move much. Can you think of a time when you felt this way? Can you think of a word that describes your body and mind when you feel slow, low, or like you don't want to play or be around others? (Offer the child the opportunity to write the word and/or draw a picture of it.)

Now let the child free-associate. You might want to prompt for the mixed pathways by saying something like, "Sometimes we might feel scared or shy or embarrassed and pretend that we're okay, when inside we're really struggling. Have you ever felt that way? What have I left out? Can you tell me any other ways your body feels?" Be patient and see where the conversation goes.

PATHWAYS TO SELF-REGULATION

Once a child understands how to observe their body's reactions to the world, they can do something profound: practice solving problems, both with you and on their own. When a child is able to recognize when their platform is getting vulnerable, a new level of self-sufficiency arises.

There are ways you can help your child reach that level. One is to show the child that you have compassion for yourself when you face a challenge or make a mistake. One way we learned is to model a range of self-awareness and respect for your internal sensations as you solve problems and face the inevitable twists and turns of being a parent. We can show our child that we can recognize and appreciate the range of how we feel as humans—the calm, in-control state; the big feelings and need-to-move state; the checked-out state; and everything in between. Then children learn that all of these are *human* states, not bad or good, the expected ways our bodies protect us. As we learned in Chapter 4, we can repair and apologize when we lose it. We can help our child learn how to regulate emotions through basic conversations, for example, asking them about their day and what they did when they felt their body shifting. In the process, we nurture our child's burgeoning self-regulation and problem-solving abilities.

Reasoning and Problem-Solving

After Mira's parents worked with her at home to help her articulate her emotions, she brought me some of the drawings she had done to give expression to her strong feelings. She acknowledged that she often had strong feelings and felt the need to move around. In particular, she loved fleeing from Leo when she returned home and he chased her around the house. When I asked her to choose a word to describe the big fight-or-flight feelings that would sometimes overwhelm her and make her feel like she was exploding, she chose *firecracker.* And her drawings, filled with red and orange swirls, illustrated the feeling. When she showed me her drawing of when she felt sad, lonely, and slow, she used the word *snail.*

The last drawing Mira showed me was of her calm, cozy, and safe place. She called it her "picnic." Her drawing showed the four members of her family, sitting on the grass, with little plates of cookies all around. She told me that she felt happiest when she was with her family at their favorite park. I noticed her parents' looks of relief at her comments as they recognized that their daughter loved her brother and that he was an important and positive part of her life (in addition, of course, to an annoyance at times).

Once a child like Mira has worked on ways to describe the different ways her platform shifts, we can help the child refine her problem-solving abilities, encouraging the child to come up with her own solutions to challenges using a more in-depth version of the resonate-and-respond technique described in Chapter 8.

Resonate. Engage with the child emotionally by doing something that lets them know we see their struggle, validates their reality, and gives them the opportunity to respond thoughtfully rather than become defensive. Sometimes, this is as simple as providing supportive silence. Then **respond** by noticing the problem/issue you observed and giving the child the opportunity to come up with multiple possible solutions to the problems they are trying to solve in order to add more flexibility to their thinking and more tools to their tool chest.

Here's an example from Mira's family that her parents used to help her solve the problem during one session:

Dad: "I saw how fast Leo grabbed your paper and ran away. Oh my!" *Then wait and see what the child says and where that lands for the child.* Do they feel witnessed? Is there a look on the child's face that tells you, "Yes, that's right; you are with me."

When her dad said that, Mira immediately resonated, saying, "Yes! All he wants to do is grab my backpack!"

Next*, encourage the child to describe their experience of what happened*, extending empathy with the goal of giving the child a chance to elaborate and let you into their thought process and to independently talk about how it feels. You can ask a simple, open-ended question, such as, "What was that like for you?" or "What did that feel like inside your body?"

If the child uses a word, or even a drawing, to describe the feeling or mood—that is, they've *named an emotion*—you've hit pay dirt. Perhaps your child will use the special word that you came up with together on the naming exercise. Any word—*mad*, *frustrated*, *happy*, *excited*, *sad*, *hot*, *embarrassed*, *scared*—tells us that the child is using emotional communication, as opposed to using the body by simply hitting or grabbing. It's okay if the word reflects raw emotion; we should help children embrace those feelings with their words.

If the child doesn't come up with a word or a description, that's okay. They might make a facial expression and then you can try to resonate with that. Or, if it feels appropriate, offer a few suggestions of emotion words: "I wonder if that made you feel frustrated? Or mad?" See how the child responds. *Remember that sometimes when we label emotions for children, it can make them feel defensive or more agitated, so tread lightly.* Then, try to help the child be *more specific* about what they're experiencing. Help them increase their vocabulary to describe their experience. For example, if they're saying, "I feel yucky," do they mean, more specifically, feeling ill, angry, frustrated, embarrassed, guilty? The more specific the words, the better the child's emotional granularity and the larger the child's emotional toolbox.

Next*, we invite the child to actively problem-solve for the future by reflecting on a question.* Say something like, "I wonder what you can do next time?" Have patience and try to help your child see things from different angles if it seems that they can think of only one solution (or a solution that you feel is suboptimal), not by telling them but by asking reflective questions such as, "Do you have any other ideas about what might help this situation?" Then wait patiently and see what the child says and if the child is open to reflection further about the problem.

Children are much more likely to implement a solution that they create for themselves. That's why it's a good idea to avoid providing or generating our *own* answers to a problem unless the child asks us for help or needs some suggestions as a scaffold to finding their own solutions. Then it's great to problem-solve together.

Mira: Using Her Mind to Solve Problems

At one session, we encouraged Mira to come up with solutions for what she could do when Leo took her things. Now that she had a way of self-regulating her emotions through the power of her own observation and the tools she created, she needed something to do for herself to regulate her emotions and feelings when she had one of her "firecracker" or "snail" experiences. In other words, she came up with her own solutions to problems. Ross Greene, the psychologist and founder of the Collaborative and Proactive Solutions (CPS) model, encourages parents to engage with their children to discover their point of view and to become collaborators in figuring out how to solve problems together. *Simply asking a child to become the problem solver and seeing their view of a problem can yield great benefits, as it did in Mira's case.*

Buoyed by the question, Mira rattled off a list of suggestions. "You mean what should I do when Leo is a firecracker?" she asked. Clearly, she was now analyzing the situation from her little brother's perspective, as well as her own. Her ideas: She could assertively try to explain to him that her backpack was *not* his property. She could find her mom or dad immediately when she came home and give one of them her backpack. Or she could leave her backpack in the car, where their mom could retrieve it later. She was generating all sorts of creative solutions, *to the delight of her parents.* Here it was, flexibility! She added more tools to her tool chest, and was now better managing her own body budget and was on her way to having greater flexibility and resilience. Then came the best part of the session: Mira spontaneously announced the plan she chose to prevent Leo from taking the drawings she had made earlier in the session. She wanted those for her wall.

This demonstrated that Mira was thinking about how to solve her sibling problems assertively and *proactively.* Her parents praised her creativity and in the weeks and months to follow, with her new tools on board and her connection with her parents as allies, the sibling conflicts reduced considerably. Not that their squabbles ceased. As I explained to her parents, the goal wasn't to *eliminate* conflict. All healthy relationships, sibling or otherwise, have a certain amount of conflict and resolution. That's how we develop and flex the muscles of social problem-solving and self-advocacy. Part of the solution involves letting the children resolve their conflicts on their own without interfering too soon.

Another part of the solution was to view each child as an

individual and see why their interactions were so complex at times. What we had discovered was that Mira and Leo both needed some practice and support to self-regulate and express themselves. Once we had helped them strengthen their individual abilities, the two of them were better equipped to resolve their conflicts on their own. Several months later, they reported that the sibling rivalry was now at a tolerable level. In fact, the two siblings had recently started to play with each other more cooperatively, and a third was on the way!

RESILIENCE-BUILDING TIP: *Play* is a powerful way to build up children's emotional literacy and pro-social problem-solving skills. We can also *model for and teach* children about the different ways our nervous system protects us by providing us with sensations we can feel. Linking sensations to basic feelings and then to emotion words helps them flexibly manage challenging situations with their growing toolbox of self-generated problem-solving skills, as they become more and more able to manage their own body budget.

10
Flourishing

> My mission in life is not merely to survive, but to thrive; and to do so with some passion, some compassion, some humor, and some style.
>
> —Maya Angelou

I became a psychologist before becoming a mother. I entered parenthood overflowing with a theoretical knowledge and five thousand hours of clinical experience. I had confidence bordering on arrogance that I possessed all the formulas I needed. But having children taught me that there are no perfect formulas for parenting, the most difficult and meaningful job I would ever have.

To be candid, I didn't always feel happy or positive as a mom early on. Sure, I felt intense waves of gratitude for my kids, my supportive husband, and my fulfilling career. But I was a mom in the trenches, often starting my day trying to talk myself into feeling happy. I was often exhausted, stretched to my limits, and doing my best to enjoy and relish parenting as it flew by, year after year. Looking back, I know that I chose that busy life, and I did the best I could.

Some parenting memories still bring a smile to my face. One of my most precious is of a warm spring day when my three daughters were all home from school because of teacher conferences. I was watching the girls take turns on a swing my husband had hung from the branches of a tall sycamore in the backyard of a home we'd recently moved into. I turned on a hose to water the soil under the tree when one of the girls rushed over, daring me to spray her. I tried and she ran away, giggling as I aimed the nozzle at her. Her two sisters joined in the game, running around in the dirt, which was quickly turning to mud. I soon realized that this was getting out of hand and we were all getting muddy.

I was typically a tidy person, not to mention that we dealt with a constant flow of laundry. But in that moment, from deep inside, my inner child emerged. Feeling a jolt of spontaneity, I shouted, "Mud bath!" I scooped up some handfuls of mud, hurling it toward my daughters. They joined in, throwing it back at me, and in no time, all four of us were covered with the dark brown mud—mud in our hair, mud coating our faces, mud between our toes. I felt like I was ten again, full of euphoria, blissfully sharing that moment with my kids.

A SURVEY OF MEMORIES

With that precious memory in mind, I wondered what moments people cherish from their own childhoods. In my work, I typically hear about problems more than happy times. After all, people don't go to psychologists to talk about what's going right. But I wanted to learn what memory traces from childhood continue to hold meaning for people years later. So I posed a question to hundreds of children and adults, from ages five to eighty-eight: What's your favorite childhood memory?

ACTIVITY: Before you read further, take a moment to answer that question yourself. What's your favorite childhood memory? Write down the first thing that comes to mind. (If you don't have a favorite, or if you don't remember anything happy from your childhood, have compassion for yourself and use *any* memory or thought that makes you feel good inside.) Let's see if yours has anything in common with the answers from my survey.

Here's a sampling from the many hundreds of people who shared favorite childhood memories:

- *"Sitting outside with my grandparents and eating big slices of watermelon. I remember the juice dripping all over me."*
- *"Spending a week on Lake Michigan with my family and running down the hot sand dunes into the cool lake."*
- *"Playing in the woods with my sister, sitting in my dad's lap, and talking to my mom for hours."*
- *"My father coming home after long trips. He used to be a pilot, and he smelled like airplane gasoline. The smell still makes me happy!"*
- *"Going to the bakery with my grandfather and getting cream puffs."*
- *"Memories of me cozied up and held tightly by my parents when walking outside in the snow."*
- *"When we spent all day picking blackberries on Orcas Island and then made jam and blackberry pie together."*
- *"During a camping trip with my family, it rained horribly the whole week, so we had to stay inside the tent trailer. My parents taught us tons of card games, we read books together, and we toasted marshmallows on the flame of a propane burner."*
- *"Sitting next to my granny, playing with the loose skin on her arm, literally bouncing it up and down! I used to do this often, cradled next to her while listening to her stories."*
- *"Summer visits at my grandparents' mountain cottage. Faint smell of moth balls on the quilts, smooth floor under my bare feet,*

crackling from the fireplace at night, fireflies out in the field, and the sound of crickets as I drifted off to sleep."

- *"When I was three or four years old, my mom would take her break (I was at YWCA day care) and come get me and we would go to the cafeteria and share a cinnamon bun."*

THE RESULTS ANALYZED

It struck me that so many replies shared common themes with each other—and with my mud bath memory. Nearly all of the recollections involved bits and pieces of sensory experiences mixed in with memories of soothing, fun, or relaxing activities. Juicy watermelon, hot sand dunes, the smell of an aviator dad's uniform, touching the skin of a grandmother's arm. And a striking formula emerged: *Nearly all of the memories were embedded in a relational experience, something shared with loved ones, as well as body-based sensory experiences and experiences that provoked safety or joy.*

To my surprise, the memories people shared reflected all of the significant points we've covered in this book: the importance of body-based experiences that form the basis of our emotional memories, and the relationships in which our experiences are embedded.

It makes sense that so many memories were about significant relationships and activities that were safe, affirming, sensory, cozy, novel, or exciting. We know that when we're supported by social connection, we take in information most effectively. In that mode, our bodies are most receptive to new experiences, so our memory holds on to moments in which we experience some degree of *novelty.* Nobody's favorite childhood memory was getting dressed and going to school every day.

Besides the prominence of human connection, many of the

recollections share another quality: coziness. Much has been written about the Danish word for coziness, *hygge*, which originates from the Norwegian word for well-being. According to Meik Wiking, the CEO of the Happiness Research Institute, *hygge* involves safety and relationships. He writes: "Time spent with others creates an atmosphere that is warm, relaxed, friendly, down-to-earth, close, comfortable, snug, and welcoming. In many ways, it is like a good hug, but without the physical contact. It is in this situation that you can be completely relaxed and yourself."

Those words also apply to my own best childhood memories: sitting under a tree playing board games with my grandmother, having tea and cookies every afternoon, sitting on my bed with her as she told me stories of her childhood. Sometimes when we were enjoying a special moment together, a satisfied look would appear on my grandmother's face and she would say a Dutch word: *gezellig*. I knew it captured something very good, but I never knew exactly what until years later, when I learned that it means "cozy or snug with other people." Clearly, many cultures agree: cozy, *hygge*, *gezellig* are feel-good words that link well-being to sharing connected moments with the people close to us.

That's also a lesson made popular in recent years by the positive psychology movement, which began with the question of how to create happiness and life satisfaction and that has also influenced the positive parenting movement.

LESSONS FROM POSITIVE PSYCHOLOGY

The field of psychology has evolved through four major waves, starting with the disease model, which focused on identifying and healing disorders of the mind (Sigmund Freud was in this early wave). Then came behaviorism, made popular in the mid-twentieth century by the psychologist B. F. Skinner, who believed that we could modify all human behaviors through the use of rewards and punishments, or consequences. The third wave was humanistic psychology, associated with psychologists Carl Rogers and Abraham Maslow, who emphasized the positive aspects of human potential, dignity, and the whole person. The fourth wave, influenced by humanism, was the positive psychology movement, which originally centered on the question of happiness and inspired a new generation of positive parenting approaches.

The psychologist Martin Seligman, considered the founder of positive psychology, initially studied happiness, thinking that was the best measure of life satisfaction. The term became popular in the late 1990s after he chose it as the theme for his term as the president of the American Psychological Association. But Seligman ran into a problem in his research. He and others discovered that when you survey people about their life satisfaction, their answers are determined by how they feel *at the moment they take the survey.* If the subject is in a cheerful mood, they rate their life satisfaction as high. If they're feeling glum at that moment, that shifts their answers downward. It became obvious that researchers weren't measuring life satisfaction but rather current mood. As I've described throughout the book, we can now view those basic feelings as reflecting our platform and balance in our body budget.

From a brain-body perspective, this phenomenon is understandable: Our bodies are constantly feeding us information, and we make sense of how good or bad we feel based on that information. Realizing the problem, Seligman shifted from studying happiness to focusing on a more all-encompassing notion: the sense of well-being.

THE MEANING OF FLOURISHING

Instead of measuring happiness or positivity, Seligman moved to studying *flourishing*, a deeper sense of well-being that includes positive emotions such as happiness but also other elements. He summed this up in his PERMA model, which stands for positive emotion, engagement, relationships, meaning, and accomplishment. The PERMA model came to be how Seligman assessed well-being. Each of these elements contributes to well-being, Seligman says, but no single one defines it. *Positive emotions*—happiness, awe, kindness, joy, empathy—contribute to well-being. So does *engagement*, being completely involved or absorbed in an activity for the sake of enjoyment. In positive psychology this state is referred to as *flow*. People may also refer to the idea of being "in the zone," when you're so engaged in an activity that you can lose track of time. The next category, *relationships*, is fundamental to well-being because relationships improve our health and, if they are positive, improve our body budget balances, as we learned in Chapter 5. Seligman found that healthy relationships help us find *meaning* and purpose in life, and to feel a sense of *accomplishment* and that what we do in our life matters. In essence, flourishing is a broad view of well-being, which includes happiness but much more.

LET'S HAVE MORE FLOURISHING

As parents, that seems an appropriate aim: not just to chase moments of happiness but to discover how we and our children can flourish and thrive. We don't accomplish this through a formula of events or by engineering experiences for our children. If you've ever tried to create a magical moment, such as a birthday party, for your child, you know this all too well. Sometimes our drive to create such experiences sets expectations too high and causes more tension and stress than joy. I learned that myself when I set up my daughter's first sleepover birthday party; she was so overwhelmed by the intensity of it all that she fled from her ten closest friends and showed up in my bedroom in tears. Sometimes our best efforts simply backfire.

To help our children flourish, we can plan less, relax more, and recognize that we don't need to work so hard at parenting. Sometimes what's best is to focus on taking good care of our children's bodies and minds—and our own—and then relish with intention the moments of joy or coziness when they happen spontaneously, launched from our well-cared-for bodies.

In my informal survey of childhood memories, so many moments reflected a sense of contentment, joy, safety, and spontaneity. Nobody mentioned costly, staged events. Instead, they described times when they felt safe, contained, and often in awe of a simple yet profound everyday experience.

The memories contained elements of flourishing, not just big emotions. They involved flow, with a sort of timeless quality. And the fact that people held on to these memories with tenderness and reverence, in some cases many decades later, reveals that they had a special meaning or a bit of novelty.

The memories in my survey had other qualities in common. Nearly all were grounded in *relational experiences*: moments with parents, siblings, cousins, grandparents, friends. They had *sensory aspects*: smelling or tasting food, seeing fireworks, feeling the grass or snow or rain. And many involved *predictable and safe patterns*: reunions at a family farm, summer vacations, holidays, rituals such as baking or lighting the candles on a menorah. These happen to be characteristics that I described throughout the book that help to build resilience.

My own parenting memory of the mud bath also involved aspects of flourishing: positive emotions (joy, spontaneity), flow (I lost track of time), and meaning. It even included a sense of accomplishment, since I was such a compulsive neat freak that indulging in a spontaneous and messy moment reflected growth on my part. In that peak "mom moment," my kids and I flourished, with laughter among us, the sun on our shoulders, and mud squishing between our toes.

LESSONS FOR THE JOURNEY

So what can we learn from reflecting on these stories of childhood? First, the importance of removing pressure from your parenting. Pressure takes away from our body budget; acceptance adds deposits. Earlier in the chapter, when I asked you to recall your favorite childhood memory, if you couldn't recall one, you aren't alone. Some of the people who answered my survey told me they had no positive childhood memories, and another subgroup shared the lone positive memory they had. If that's your experience, too, I hope that the message and information I have shared in this book—the power to create new experiences—bring you some comfort. In my career I've witnessed so many lives rebuilt after people experienced trauma and unimaginable tragedies. I've gained a profound appreciation for the human capacity to heal and move forward. While I'm realistic about the toll being a parent takes, especially after a global pandemic, I remain hopeful.

I also hope reading this book has given you a deeper understanding and sense of compassion for yourself as a parent. I hope you feel more confident in being yourself as a parent, doing what you need to do, not what you *think you should be doing* or what you see other parents doing. The interactions that build resilience for our children should also feel nourishing to you. Your well-being matters.

And consider these suggestions to increase the coziness or comfort factor in your family's life, helping you and your children to flourish. This isn't meant to be a to-do list but rather a series of suggestions you can customize to your family, your children, your unique situation.

- **ELEVATE RELATIONSHIPS AS A TOP PRIORITY:** *Realize how important other people are in your life and in your child's life. We are social creatures, and even though it may feel like cultivating and nurturing relationships takes more energy than you have, what you gain from staying in contact with those you love provides nurturing energy in the long run. It benefits you and your children to have safe, fun, and predictable relational experiences on a regular basis. Try to involve family, grandparents, aunts and uncles, and close friends in your life—if you deem that they are a positive and healthy influence, of course. If you have quarrels or difficulties with your parents, siblings, or devoted friends, do what you can (within healthy boundaries) to try to heal old wounds, because relationships and a sense of togetherness are the best ally you and your child have.*

- **GO FOR FLOW:** *Remember that flow, and not necessarily "happiness," is a driver of flourishing, so occasionally give yourself permission to do something with your child that makes you lose track of time, even if it's for a few minutes. Let your guard down, allow yourself to break the rules, and let go of your never-ending to-do list in the service of spontaneity and following your child's lead. Doing so might lead to a treasured memory. You never know until you experience the flow of engagement.*

- **FIND MOMENTS OF CONNECTION DURING EVERYDAY EXPERIENCES:** *You don't have to create new and elaborate experiences for your child to have moments of connection with you. Everyday experiences such as mealtimes and bath times provide natural opportunities for comfort. Sure, mealtimes can feel like a chore, with the planning, the preparation, and the clean-up afterward. But they're also opportunities to have the sensory experiences of smell, taste, listening, and seeing paired with your loving and engaged presence, so that your child associates pleasure* *with mealtimes. Rather than focusing on the end product, eating, slow it down and enjoy the meal together. Turn on some quiet music (if that makes you feel good) or have a spontaneous picnic on the floor with your child, rather than eating later once they've been fed. Savor each bite as you savor the essence of the child sitting in front of you.*

If your child is old enough, consider deepening the experience by asking them to tell you one thing they remember about the day, or something that made them feel good or *uncomfortable that day, giving equal weight or importance to each one. This tells the child you aren't interested only in their positive emotions but the negative ones, too, which are just as important. If your child doesn't want to talk, that's okay. Just stay present and open to the moment. Being in space together in the safety of nonjudgment is powerful in itself and doesn't have to include talking.*

- **ADD A BIT OF FUN, SAFE UNPREDICTABILITY NOW AND THEN:** *While humans love knowing what to expect, we remember novelty more. So if you're up for it, add some fun to your child's day. Maybe prop up your three-year-old's teddy bear in the kitchen holding an apple, and watch the child discover it there in the morning. Novelty! "Now what was your Teddy up to last night?" Maybe mix green food coloring into your six-year-old's pancake mix and watch the expression on their face when you serve the pancakes with a smile. (Of course, in deference to individual differences, if your child frowns or isn't impressed with such little tricks, that's a perfect time to co-regulate with love and tenderness.) Remember to respect each child's perception of the world and to add novelty that the child perceives as safe or joyful.*

- **UP THE COZINESS FACTOR:** *As we've seen, there's a universal appreciation for comfort and connection that leads to feeling cozy. You'll have to discover what feels comfortable and cozy to your particular child and family situation, whether it's snuggling under* *warm blankets, enjoying a bowl of delicious soup, sitting by a crackling fire, or lying on a sidewalk or rooftop watching airplanes or cloud formations. You'll feel it when it happens, as the feeling arises from safety and connection. The opportunities are there, ready for the taking, and require not hours but only minutes or even seconds.*

How can busy, overscheduled parents pull this off? By noticing and being present—difficult as that may be while parenting your children. If that feels impossible (as it often did for me when my kids were young), that's okay. Self-compassion is our ally, our own cozy warm blanket to ease the self-doubt that all parents occasionally experience. *We all do the best we can with the resources we have in each moment and each day. That's good enough.*

We end where we began, with relationships. Nothing is more comforting to a child than a parent's calm and loving presence. It's parents and caregivers who promote a child's felt sense of safety in the body. We customize our parenting to each child's body and brain according to how that child experiences safety. We use this knowledge to help our children move through their challenge zones, providing the backbone and presence that shows them they can gain new strengths and eventually develop the self-regulation and self-sufficiency to thrive on their own terms.

We parents are on a hero's journey. It's an incredibly challenging and high-pressure journey because the stakes are so high. But if you can stay grounded in compassion when you feel yourself fall short, you can provide that cozy, comforting blanket for yourself and your children. I wish you many such moments of warmth and connection with your family, and a lifetime of well-being.

ACKNOWLEDGMENTS

I owe a debt of gratitude to the many people whose lives have touched mine and led to the birth of this book. First, I'm so grateful for all of the children, parents, and families I've had the privilege of walking alongside. I've learned about hope, courage, and love from you all, and you taught me more about relationships and resilience than anything I ever learned in school.

Nobody has supported me more than my mom. The hours we spent on the phone while I was writing this book gave me strength and confirmed my belief that these ideas were worth sharing with parents. This book is dedicated to her.

Tom Fields-Meyer, my friend, wise guide, and longtime editor, gives flight to my ideas like nobody else. It's a joy to have an editor who shares a common vision of a world with more tolerance and compassion for children. My first writing coach, Teresa Barker, gently incubated my dream while teaching me about the publishing process. Leah Caldwell, my technical editor, skillfully organized my three hundred–plus endnotes and endless bibliography and made it seem easy.

My literary agent, Amy Hughes, who "got me" (and this book) from day one, is a constant source of inspiration, guidance, support, and encouragement. Words can't express my gratitude for her wisdom and concern, not only for this book but for me as a person and a professional.

The amazing team at PESI and PESI Publishing, led by Linda Jackson and Karsyn Morse, published my first two books. *Beyond Behaviors* was the launchpad for *Brain-Body Parenting*, and the PESI organization feels like family to me. Zachary Taylor and Livia Kent at Psychotherapy Networker have been so supportive of my work, and the field of mental health is benefiting from your courageous leadership.

I am delighted that *Brain-Body Parenting* landed in the capable hands of the team at Harper Wave. It was a joy to work with my editor, Julie Will, and my trail guide and editorial assistant Emma Kupor, whose responsiveness made this process smooth and seamless. I'm indebted to Karen Rinaldi, Harper Wave publisher; David Koral, senior production editor; Brian Perrin and Laura Cole in marketing; Leslie Cohen in publicity; Elina Cohen, the interior designer; and Milan Bozic, the jacket designer. Every member of the team contributed to the final product.

I'm very grateful for the strategic help and moral support from my team at Zilker Media: Nichole Williamson, Sydney Panter, Macey Pieterse, and Melanie Cloth.

I learned how to do the work of relationship-based practice from Serena Wieder, whose lifework has left an indelible imprint on how I support children and families. I'll never forget our days in the "bunker" as I learned from her and from many others, including Ricki Robinson and Gil Foley. I'm grateful to Monica Osgood and the entire faculty and staff of the Profectum Foundation for their continual support and for their work in bringing relationship-based practices to the world.

I will always be grateful to Dr. Stephen Porges, whose insights about the autonomic nervous system are changing lives and paradigms for the better. It's no wonder his work is often referred to as the "neurobiology of love." I'm appreciative of the time he has taken over the years to read my work, offer suggestions, and for his support of my translation of his theories into clinical practice and parenting. And I so enjoy doing Zoom presentations together!

By the most incredible stroke of luck, the professionals who constitute the "dream teams" of my practice (that I would have traveled across the country to consult with) live in my own backyard of Los Angeles. I'm grateful to trailblazer Connie Lillas for her lifework on the neurorelational framework, and for her bold style and courage to bring neuroscience into clinical practice. Dr. Marilyn Mehr helped me find my assertive voice while working in a medical setting early on in my career. When I began my journey into the world of development and infant mental health, Marie Kanne Poulsen generously opened the doors to training at Children's Hospital Los Angeles. Diane Danis, developmental pediatrician, guided many of our teams while helping us understand the brain *and* body of each child we treated together. Amy Johnson, SLP; Susan Spitzer, OTR/L; Kate Crowley, OTR/L; Margaret Mortimore, PT; and Uyen Nguyen, OTR/L, were on my speed dial as we worked together with so many families over the past several decades. I'll always be grateful for the excellent relationship-based agencies in our area, led by my valued colleagues Diane Cullinane, Andrea Davis, and Ben Russell. I'm so thankful for my colleague Tina Payne Bryson, another shining star of Pasadena, California, and the world. I so enjoyed our lunches and dinners when she was building the multidisciplinary team practice in Pasadena. It was an honor to present to the incredible staff at the Center for Connection when *Beyond Behaviors* came out. Recently, my "interoception group" friends Ira, Kelly, Annemarie, Amanda, Nanci, Renee, and Nina have been a source of inspiration and intellectual nourishment. What fun it is to be talking about the topic that is changing the face of how we view emotional processing and self-regulation!

I'm fortunate to have siblings and their partners, Glenn, Warren, Hank, Ana, KK, and Kevin, whose love I can feel no matter the time or distance that separates us. Glenn, you are my right hand and heart; Hank, you carry on the family legacy of generosity; and, KK, nobody will ever share our special history and those magical early years we had together. I'm grateful for my "daddy" Doug, and mom-in-law, Jane, who always asked about how work was going, and for the extended Delahooke family. I'm also grateful and crazy about all my nieces and nephews, who are launching into their own unique journeys as adults.

I wouldn't have had the strength to do this work without the loving care and concern of my inner circle of loved ones. These are the folks whose dinners and lunches, phone conversations, texts, frog gifts, dried fruit, and friendships have sustained me over the decades: Mira, Vincent, Joe, Lisa, Cindy, Garth, Jaime, Jordan, Brigitta, Olga, Mike, Stephen, Sally, Terry, Susan, Beau, Karen, Ron, Leeann, Diane, Fred, Bill, Carmen, Kathy, Ernesto, Jeanne-Anne, Yatidi, Mary, Ray, Lourdes, Lycke, and Hanneke.

To my daughters, Nikki, Kendra, and Alexa; and son-in-law Tyler. You are my best teachers, and I continue to learn from each of you. Thank you for being the courageous people you are and for continuing to help me be the best mom I can be, even when it's uncomfortable.

To my love, Scott, forever partner, co-parent, and trusty Sherpa, the one who always helps me find my way when I get lost in hotel lobbies. *This book belongs to both of us.*

And finally, I'm so grateful to the little lightbulb of my life, Skyler. Oma loves you so much. Let's keep playing and playing.

GLOSSARY

AUTONOMIC NERVOUS SYSTEM (ANS): The part of the nervous system that regulates the internal organs in the body outside of our awareness. The system contains two divisions, the sympathetic and the parasympathetic branches.

BLUE PATHWAY: The term I use to describe the dorsal vagal pathway of the parasympathetic branch of the autonomic nervous system.

BODY BUDGET: A term (coined by Dr. Lisa Feldman Barrett) describing how your brain allocates energy resources within your body; the scientific term is *allostasis*.

EMOTIONAL GRANULARITY: A term that describes the ability to construct more or less specific emotional experiences and perceptions (Barrett, 2017).

GREEN PATHWAY: The term I use to describe the ventral vagal pathway of the parasympathetic branch of the autonomic nervous system. It is also known as the social engagement system in the Polyvagal theory.

INTEROCEPTION: Internal sensations that provide information from your body's organs, tissues, hormones, and immune system (Barrett, 2017).

INTEROCEPTIVE AWARENESS: When you perceive or become aware of sensations from deep inside your body.

NEUROCEPTION: Dr. Stephen Porges's term for the subconscious detection of safety and threat (*see also* safety system or safety detection system).

PLATFORM: Shorthand for the brain-body connection, or the physiological state of our body. We are never just a "body" or a "brain"; we are always both. The platform is influenced by the state of the autonomic nervous system.

POLYVAGAL THEORY: Introduced in 1994 by Dr. Stephen Porges, this theory links the evolution of the mammalian autonomic nervous system to social behaviors. The main premise of the theory is that human beings need safety and that our biology is protectively positioned to keep us safe.

RED PATHWAY: The term I use to describe the sympathetic, or fight-or-flight, pathway of the autonomic nervous system.

SAFETY SYSTEM OR SAFETY DETECTION SYSTEM: Simplified terms for neuroception.

NOTES

Chapter 1: How to Understand Your Child's Physiology—and Why It's Important

15 "We are the caretakers of each other's nervous systems as much as we are the caretakers for our own": Barrett, "Bloom Where You Are Rooted: What Neuroscience Can Teach Us about Harnessing Passion and Productivity."

18 keeping us safe: Porges, "Neuroception: A Subconscious System for Detecting Threats and Safety," 19.

18 constantly predicting what's going on: Barrett, *How Emotions Are Made.*

18 Our brain and body are constantly talking to each other: Porges, *The Polyvagal Theory: Neurophysiological Foundations of Emotions, Attachment, Communication, and Self-Regulation*; Craig, *How Do You Feel? An Interoceptive Moment with Your Neurobiological Self.*

19 unique two-way system of communication between the body and brain: Porges and Dana, *Clinical Applications of the Polyvagal Theory: The Emergence of Polyvagal-Informed Therapies*; Porges, *The Polyvagal Theory: Neurophysiological Foundations of Emotions, Attachment, Communication, and Self-Regulation.*

19 *our nervous system, serves as a neural platform that influences human behaviors*: Porges and Furman, "The Early Development of the Autonomic Nervous System Provides a Neural Platform for Social Behavior: A Polyvagal Perspective," 106.

19 continuum from receptive to defensive: Porges, *The Polyvagal Theory: Neurophysiological Foundations of Emotions, Attachment, Communication, and Self-Regulation.*

19 When we're feeling safe, we're in a receptive mode: Porges, *The Polyvagal Theory: Neurophysiological Foundations of Emotions, Attachment, Communication, and Self-Regulation*.

19 state of their autonomic nervous system: Porges and Furman, "The Early Development of the Autonomic Nervous System Provides a Neural Platform for Social Behavior: A Polyvagal Perspective."

20 what's called *allostasis*: McEwen, "Protective and Damaging Effects of Stress Mediators."

20 *body budgeting*: Barrett, *How Emotions Are Made*.

20 resources like water, salt, and glucose as you gain and lose them: Barrett, *Seven and a Half Lessons about the Brain*, 6.

20 deposits or withdrawals in our *body budget*: Barrett, *Seven and a Half Lessons about the Brain*; Barrett, *How Emotions Are Made*. All forward references to the body budget are attributed to Dr. Barrett.

22 changes that require an adjustment or response: Cleveland Clinic, "Stress."

22 power of human adaptation: Porges, "The Polyvagal Theory: New Insights into Adaptive Reactions of the Autonomic Nervous System."

23 *just-right challenge*: Schaaf and Miller, "Occupational Therapy Using a Sensory Integrative Approach for Children with Developmental Disabilities"; Ayres, *Sensory Integration and the Child: Understanding Hidden Sensory Challenges*.

23 *challenge zone*: The general idea is from the zone of proximal development. Vygotsky, *Mind in Society: The Development of Higher Psychological Processes*.

24 *frustration tolerance*: APA Dictionary of Psychology, "Frustration Tolerance."

24 can delay gratification: APA Dictionary of Psychology, "Frustration Tolerance."

29 *allostatic load*: McEwen and Norton Lasley, *The End of Stress as We Know It*.

31 *nurture our child's nature*: Immordino-Yang et al., "Nurturing Nature: How Brain Development Is Inherently Social and Emotional, and What This Means for Education."

Chapter 2: Neuroception and the Quest to Feel Safe and Loved

33 "Before they can make any kind of lasting change at all in their behavior, they need to feel safe and loved": Perry and Szalavitz, *The Boy Who Was Raised as a Dog*, 273.

34 humans are on a quest to feel safe: Porges, "Neuroception: A Subconscious System for Detecting Threats and Safety."

34 *executive functions*: Diamond, "Executive Functions."

35 maintaining allostasis: McEwen and Norton Lasley, *The End of Stress as We Know It.*

35 Our nervous system: Stanfield, *Principles of Human Physiology*; McEwen and Norton Lasley, *The End of Stress as We Know It.*

35 a remarkable monitoring system: Porges, "Neuroception: A Subconscious System for Detecting Threats and Safety."

36 that detects safety or threat: Porges, *The Polyvagal Theory: Neurophysiological Foundations of Emotions, Attachment, Communication, and Self-Regulation.*

36 neuroception: Porges, "Neuroception: A Subconscious System for Detecting Threats and Safety."

36 automatically and subconsciously: Porges, "Neuroception: A Subconscious System for Detecting Threats and Safety."

36 *interoception*: Porges, "The Infant's Sixth Sense: Awareness and Regulation of Bodily Processes"; Craig, *How Do You Feel? An Interoceptive Moment with Your Neurobiological Self.*

36 lightning-fast prediction machine: Barrett, *How Emotions Are Made.*

37 "Your brain is always predicting, and its most important mission is predicting your body's energy needs, so you can stay alive and well": Barrett, *How Emotions Are Made*, 66.

37 tells the body what to do quickly and most efficiently: Porges, *The Polyvagal Theory: Neurophysiological Foundations of Emotions, Attachment, Communication, and Self-Regulation.*

38 our body's subconscious surveillance system: Porges, "Neuroception: A Subconscious System for Detecting Threats and Safety."

39 sense of unease into *perception*: Dana, *The Polyvagal Theory in Therapy: Engaging the Rhythm of Regulation.*

40 Past experiences influence children and help them predict: Barrett, *How Emotions Are Made.*

41 a tribute to the body's survival instincts: Porges, *The Polyvagal Theory: Neurophysiological Foundations of Emotions, Attachment, Communication, and Self-Regulation.*

41 remind the nervous system of something the safety-detection system initially coded as threatening: Porges, "Neuroception: A Subconscious System for Detecting Threats and Safety"; Barrett, *How Emotions Are Made.*

43 We need a certain amount of stress for the brain to realize: Perry and Szalavitz, *The Boy Who Was Raised as a Dog.*

43 Stress that is *predictable, moderate, and controlled*: Perry and Szalavitz, *The Boy Who Was Raised as a Dog*, 314.

43 when stress is *unpredictable, severe, and prolonged*: Perry and Szalavitz, *The Boy Who Was Raised as a Dog*, 314.

43 the body budget can run a deficit: Barrett, *Seven and a Half Lessons about the Brain*; Barrett, *How Emotions Are Made.*

43 allostatic load: McEwen, "Protective and Damaging Effects of Stress Mediators."

43 can contribute to diseases with heavy links to stress: McEwen and Norton Lasley, *The End of Stress as We Know It.*

44 the most important way for all humans to maintain their body budgets is to get enough sleep: Goldstein and Walker, "The Role of Sleep in Emotional Brain Function."

44 Sleep sets the foundation: Walker, *Why We Sleep: Unlocking the Power of Sleep and Dreams.*

47 influenced by our past and present experiences: Barrett, *How Emotions Are Made.*

48 *resolve or reduce* the cues of perceived threat that are triggering the child (if possible and appropriate to the situation), and (2) *bring in* cues of safety: Dana, *The Polyvagal Theory in Therapy: Engaging the Rhythm of Regulation*, 42.

51 a child with a frown could be angry, frustrated, or concentrating—or be experiencing some other emotion or sensation: Barrett, *How Emotions Are Made.*

55 Neuroception: Porges, "Neuroception: A Subconscious System for Detecting Threats and Safety."

Chapter 3: The Three Pathways and the Check-In

57 "Our body is always doing what it thinks is best for us": Porges, "Polyvagal Theory and Regulating Our Bodily State."

57 *body and brain interact with each other*: Price and Hooven, "Interoceptive Awareness Skills for Emotion Regulation: Theory and Approach of Mindful Awareness in Body-Oriented Therapy (MABT)"; Porges and Dana, *Clinical Applications of the Polyvagal Theory: The Emergence of Polyvagal-Informed Therapies.*

59 reading the internal and external environment and interactions with other people: Porges, *The Polyvagal Theory: Neurophysiological Foundations of Emotions, Attachment, Communication, and Self-Regulation.*

59 how you feel ("affect") has two main features: the *feeling of pleasant or unpleasant* (known as "valence") and the degree of *calmness or agitation* (known as "arousal"): Posner, Russell, and Peterson, "The Circumplex Model of Affect: An Integrative Approach to Affective Neuroscience, Cognitive Development, and Psychopathology"; Russell and Barrett, "Core Affect, Prototypical Emotional Episodes, and Other Things Called Emotion: Dissecting the Elephant."

59 "always some combination of valence and arousal": Barrett, *How Emotions Are Made*, 74.

60 *balance their body budget through our loving interactions*: Barrett, *How Emotions Are Made.*

60 *central nervous system*: Stanfield, *Principles of Human Physiology.*

60 *peripheral nervous system*, which contains the *somatic nervous system*—involved in the movement of our skeletal muscles: Stanfield, *Principles of Human Physiology.*

60 homeostasis: Stanfield, *Principles of Human Physiology.*

60 *the sympathetic and the parasympathetic*, which have different effects on our organs: Stanfield, *Principles of Human Physiology.*

61 the Polyvagal theory: Porges, *The Polyvagal Theory: Neurophysiological Foundations of Emotions, Attachment, Communication, and Self-Regulation.*

61 *sympathetic nervous system*: Porges, *The Polyvagal Theory: Neurophysiological Foundations of Emotions, Attachment, Communication, and Self-Regulation.*

62 make inferences based on their tone of voice, muscle movements, heart and lung activity, body gestures, and behaviors: Porges, *The Polyvagal Theory: Neurophysiological Foundations of Emotions, Attachment, Communication, and Self-Regulation*; Lillas and Turnbull, *Infant/Child Mental Health, Early Intervention, and Relationship-Based Therapies: A Neurorelational Framework for Interdisciplinary Practice.*

62 behaviors that cluster together, providing clues about the state of a child's physiology: Lillas and Turnbull, *Infant/Child Mental Health, Early Intervention, and Relationship-Based Therapies: A Neurorelational Framework for Interdisciplinary Practice*. These clusters of behaviors were first described as states of consciousness by the early pioneers of infant research, including

Kathryn Barnard, T. Berry Brazelton, Georgia DeGangi, and Porges, and are directly influenced by the autonomic nervous system and reflect responses to internal and external sensory stimuli. The markers for these clusters of behaviors have been refined subsequently by Lillas. See Lillas, "Handouts."

62 *wearable sensors*: Taj-Eldin et al., "A Review of Wearable Solutions for Physiological and Emotional Monitoring for Use by People with Autism Spectrum Disorder and Their Caregivers."

63 the first clinically validated and FDA-approved device that provides valuable insights for individuals with epilepsy and their caregivers: "Embrace2."

63 The Green Pathway: The colors and descriptions of the autonomic pathways are adapted from Lillas and Turnbull, *Infant/Child Mental Health, Early Intervention, and Relationship-Based Therapies: A Neurorelational Framework for Interdisciplinary Practice.*

63 a person feels safe and social, connected with others and with the world around them: Dana, *The Polyvagal Theory in Therapy: Engaging the Rhythm of Regulation.*

63 we send others signals of connection and communication: Dana, *The Polyvagal Theory in Therapy: Engaging the Rhythm of Regulation.*

64 In our *physical bodies*, we might see: Lillas and Turnbull, *Infant/Child Mental Health, Early Intervention, and Relationship-Based Therapies: A Neurorelational Framework for Interdisciplinary Practice.*

65 "gleam in the eye": Greenspan, "The Greenspan Floortime Approach: Interaction."

66 cycle in and out of each pathway: Dana, *The Polyvagal Theory in Therapy: Engaging the Rhythm of Regulation.*

66 more costly to the body budget: Barrett, *How Emotions Are Made.*

67 *biobehavioral* reaction: Dana, *The Polyvagal Theory in Therapy: Engaging the Rhythm of Regulation.*

67 feel the need to *move*: Dana, *The Polyvagal Theory in Therapy: Engaging the Rhythm of Regulation.*

67 When our nervous system detects threat: Porges, *The Pocket Guide to the Polyvagal Theory: The Transformative Power of Feeling Safe.*

67 In our *physical bodies*, we might see: Lillas and Turnbull, *Infant/Child Mental Health, Early Intervention, and Relationship-Based Therapies: A Neurorelational Framework for Interdisciplinary Practice.*

68 *The Body Keeps the Score*: Van der Kolk, *The Body Keeps the Score: Brain, Mind, and Body in the Healing of Trauma.*

69 The red pathway is the pathway of "mobilization": Porges, *The Polyvagal*

Theory: Neurophysiological Foundations of Emotions, Attachment, Communication, and Self-Regulation.

69 Hence the phrase "fight or flight": Porges, *The Polyvagal Theory: Neurophysiological Foundations of Emotions, Attachment, Communication, and Self-Regulation.*

70 helping the child to protect themselves through *instinctive* rather than *willful* or rude behaviors: Porges, "Neuroception: A Subconscious System for Detecting Threats and Safety."

71 distinguish the sounds of the human voice: Porges, "Stephen Porges (Polyvagal Perspective and Sound Sensitivity Research)."

71 middle ear muscles shift away from distinguishing the nuances of human voices: Porges, "Stephen Porges (Polyvagal Perspective and Sound Sensitivity Research)."

71 When a child is in the red pathway, they might register a neutral facial expression as an angry one: Porges, "Neuroception: A Subconscious System for Detecting Threats and Safety"; Porges and Furman, "The Early Development of the Autonomic Nervous System Provides a Neural Platform for Social Behavior: A Polyvagal Perspective."

72 a person conserves energy by withdrawing from connection: Dana, *The Polyvagal Theory in Therapy: Engaging the Rhythm of Regulation.*

73 In our *physical bodies*, we might see: Lillas and Turnbull, *Infant/Child Mental Health, Early Intervention, and Relationship-Based Therapies: A Neurorelational Framework for Interdisciplinary Practice.*

73 nervous system is detecting very high levels of threat and is protectively conserving energy: Dana, *The Polyvagal Theory in Therapy: Engaging the Rhythm of Regulation.*

75 three states of the autonomic nervous system: Porges, "The Polyvagal Theory: New Insights into Adaptive Reactions of the Autonomic Nervous System."

75 the various pathways blend or overlap: Berntson and Cacioppo, "Heart Rate Variability: Stress and Psychiatric Conditions"; Christensen et al., "Diverse Autonomic Nervous System Stress Response Patterns in Childhood Sensory Modulation."

75 meditation is an example of stillness without fear: Dana, *The Polyvagal Theory in Therapy: Engaging the Rhythm of Regulation.*

75 likely involves a blended state: Porges, "Reciprocal Influences between Body and Brain in the Perception and Expression of Affect: A Polyvagal Perspective"; Dana, *The Polyvagal Theory in Therapy: Engaging the Rhythm of Regulation.*

76 the brain's main job is to maintain the body budget: Barrett, *How Emotions Are Made.*

78 either of which would be metabolically costly: Barrett, *How Emotions Are Made.*

79 *how often, how intense, and how prolonged*: McEwen and Norton Lasley, *The End of Stress as We Know It*; Perry and Szalavitz, *The Boy Who Was Raised as a Dog.*

79 weekly journal: Adapted from Lillas and Turnbull, *Infant/Child Mental Health, Early Intervention, and Relationship-Based Therapies: A Neurorelational Framework for Interdisciplinary Practice.*

79 level of distress from 1 to 5: Adapted from Joseph Wolpe, who developed the Subjective Units of Distress Scale, or SUDS. See Wolpe, *The Practice of Behavior Therapy*; and Lillas, "NRF Foundations Manual" (forthcoming).

80 about 30 percent of the time: This guideline is based upon clinical consensus in private conversation with Dr. Connie Lillas, based on our combined years of clinical experience. We support the need for research in this arena to verify or contrast this suggestion.

82 the child senses safety or threat first on a *nonverbal* level: Porges, "Neuroception: A Subconscious System for Detecting Threats and Safety."

Chapter 4: Nurturing Children's Ability to Self-Regulate

93 "To connect and to co-regulate with others is our biological imperative": Porges, *The Pocket Guide to the Polyvagal Theory: The Transformative Power of Feeling Safe*, 51.

93 they have a strong foundation for building resilience: Porges and Furman, "The Early Development of the Autonomic Nervous System Provides a Neural Platform for Social Behavior: A Polyvagal Perspective."

93 attuned parenting helps build the brain architecture: Harvard University, Center on the Developing Child, "Resilience."

95 intentional control (regulation) of one's thoughts, emotions, and behaviors: McClelland et al., "Self-Regulation: The Integration of Cognition and Emotion."

95 manage how we act and feel: Mahler, *Interoception: The Eighth Sensory System.*

96 do better academically and socially: Duckworth and Carlson, "Self-Regulation and School Success"; Montroy et al., "The Development of Self-Regulation across Early Childhood"; Artuch-Garde et al.,

"Relationship between Resilience and Self-Regulation: A Study of Spanish Youth at Risk of Social Exclusion."

96 the expectation gap: Zero to Three, "Parent Survey Reveals Expectation Gap for Parents of Young Children."

97 continues into early adulthood: Harvard University, Center on the Developing Child, "The Science of Adult Capabilities."

97 nourishes children's growing capacity to self-regulate: Tronick, *The Neurobehavioral and Social-Emotional Development of Infants and Children*; Tronick and Beeghly, "Infants' Meaning-Making and the Development of Mental Health Problems"; Feldman, "The Adaptive Human Parental Brain: Implications for Children's Social Development"; Feldman, "Infant–Mother and Infant–Father Synchrony: The Coregulation of Positive Arousal"; Fogel, *Developing through Relationships: Origins of Communication, Self, and Culture.*

98 *regulate their body budget*: Barrett, *How Emotions Are Made.*

98 as we notice a baby's physical needs: Zero to Three, "It Takes Two: The Role of Co-Regulation in Building Self-Regulation Skills."

98 *responsive* interactions: Zero to Three, "It Takes Two: The Role of Co-Regulation in Building Self-Regulation Skills."

98 "a shared state of calmness results when we reframe another's behavior and identify and reduce their stresses": Shanker, *Reframed: Self-Reg for a Just Society*, 274.

98 may be more vulnerable physiologically: Bush et al., "Effects of Pre- and Postnatal Maternal Stress on Infant Temperament and Autonomic Nervous System Reactivity and Regulation in a Diverse, Low-Income Population"; Gray et al., "Thinking across Generations: Unique Contributions of Maternal Early Life and Prenatal Stress to Infant Physiology."

99 our brain detects: Porges, *The Polyvagal Theory: Neurophysiological Foundations of Emotions, Attachment, Communication, and Self-Regulation.*

100 a cornerstone concept: Fogel, *Developing through Relationships: Origins of Communication, Self, and Culture*; Tronick, *The Neurobehavioral and Social-Emotional Development of Infants and Children*; Zero to Three, "It Takes Two: The Role of Co-Regulation in Building Self-Regulation Skills."

100 *mutual regulation*: Tronick, *The Neurobehavioral and Social-Emotional Development of Infants and Children.*

101 Only about 30 percent of mother-baby interactions are well matched

or coordinated on the first pass: Tronick, "Emotions and Emotional Communication in Infants."

101 "messiness is an inherent quality of infant–caregiver interactions, and therefore the task of creating shared meanings is a daunting one for infants, children, and adults alike": Tronick and Beeghly, "Infants' Meaning-Making and the Development of Mental Health Problems," 7.

102 *repair process*: Tronick, *The Neurobehavioral and Social-Emotional Development of Infants and Children.*

104 develop hardiness and grit: Duckworth, *Grit: The Power of Passion and Perseverance.*

104 *only* the mismatches and ruptures, and not the repairs: Perry and Szalavitz, *The Boy Who Was Raised as a Dog.*

104 stress becomes toxic or traumatic when the child doesn't have adults in the picture who can offer support: Perry and Szalavitz, *The Boy Who Was Raised as a Dog.*

105 *without tolerable levels of stress*: Perry and Szalavitz, *The Boy Who Was Raised as a Dog.*

105 "if moderate, predictable and patterned, it is stress that makes a system stronger and more functionally capable": Perry and Szalavitz, *The Boy Who Was Raised as a Dog*, 40.

106 *frequency, intensity, and duration*: Perry and Szalavitz, *The Boy Who Was Raised as a Dog.*

108 Italian researchers: Di Pellegrino et al., "Understanding Motor Events: A Neurophysiological Study"; Iacoboni, *Mirroring People: The Science of Empathy and How We Connect with Others.*

108 a mother's face naturally "mirrors" her baby's feelings: Winnicott, "Mirror-Role of Mother and Family in Child Development."

108 experiences changes in her *own* nervous system: Ebisch et al., "Mother and Child in Synchrony: Thermal Facial Imprints of Autonomic Contagion."

114 "serve-and-return" interactions: Harvard University, Center on the Developing Child, "Serve and Return."

114 based on the other player's anticipated actions: Fogel, *Developing through Relationships: Origins of Communication, Self, and Culture.*

114 "The energy of reciprocity is one of sending care back and forth": Dana, *The Polyvagal Theory in Therapy: Engaging the Rhythm of Regulation*, 125.

115 *use their words or gestures*: Greenspan and Breslau Lewis, *Building Healthy Minds: The Six Experiences That Create Intelligence and Emotional Growth in Babies and Young Children.*

116 Soft eyes: Kendo Notes Blog, "'Soft Eyes,' a Way of Seeing and Being–Quotes and Resources."

120 "Kids do well if they can": Greene, *The Explosive Child.*

Chapter 5: Taking Care of Yourself

123 "Nurturing your own development isn't selfish. It's actually a great gift to other people": Hanson, *Buddha's Brain: The Practical Neuroscience of Happiness, Love, and Wisdom.*

126 the negativity bias: Rozin and Royzman, "Negativity Bias, Negativity Dominance, and Contagion."

127 Bad experiences stick to us like Velcro: Hanson, *Hardwiring Happiness: The New Brain Science of Contentment, Calm, and Confidence.*

127 negative experiences are simply more "sticky": Hanson, *Hardwiring Happiness: The New Brain Science of Contentment, Calm, and Confidence.*

128 slow down cellular aging: Epel et al., "Can Meditation Slow Rate of Cellular Aging? Cognitive Stress, Mindfulness, and Telomeres."

128 *From Neurons to Neighborhoods: The Science of Early Childhood Development*, which translated science into practice: Shonkoff and Phillips, eds., *From Neurons to Neighborhoods: The Science of Early Childhood Development.*

128 *Vibrant and Healthy Kids*: National Academies of Sciences, Engineering, and Medicine, *Vibrant and Healthy Kids: Aligning Science, Practice, and Policy to Advance Health Equity.*

128 "Ensuring the well-being of caregivers by supporting and caring for them is critical for healthy child development": National Academies of Sciences, Engineering, and Medicine, *Vibrant and Healthy Kids: Aligning Science, Practice, and Policy to Advance Health Equity*, 27.

129 "undeniable effects of racism in our society": Associated Press, "Fauci Says Pandemic Exposed 'Undeniable Effects of Racism.'"

132 hydrating yourself: Mayo Clinic, "Nutrition and Healthy Eating."

133 "Sleep is a non-negotiable biological necessity. It is your life support system": Walker, "Sleep Is Your Superpower."

133 Sleep supports every system of our body: Walker, *Why We Sleep: Unlocking the Power of Sleep and Dreams.*

134 extended family groups that were the norm: Perry and Szalavitz, *The Boy Who Was Raised as a Dog.*

134 *it's important not to let sleep deprivation become chronic*: Walker, *Why We Sleep: Unlocking the Power of Sleep and Dreams.*

135 we survive and thrive through human connection: Luthar and Ciciolla, "Who Mothers Mommy? Factors That Contribute to Mothers' Well-Being"; Perry and Szalavitz, *The Boy Who Was Raised as a Dog.*

135 feel cared for as parents: National Academies of Sciences, Engineering, and Medicine, *Vibrant and Healthy Kids: Aligning Science, Practice, and Policy to Advance Health Equity*; Luthar and Ciciolla, "Who Mothers Mommy? Factors That Contribute to Mothers' Well-Being."

135 "*Who mothers mommy?*": Luthar and Ciciolla, "Who Mothers Mommy? Factors That Contribute to Mothers' Well-Being."

135 loneliness and social isolation: Cacioppo and Cacioppo, "The Growing Problem of Loneliness."

135 A large study conducted by Cigna: Cigna, "Cigna U.S. Loneliness Index."

136 moments of connection with another caring adult: Cacioppo and Cacioppo, "The Growing Problem of Loneliness."

139 positive impact of mindfulness: Baer, Lykins, and Peters, "Mindfulness and Self-Compassion as Predictors of Psychological Wellbeing in Long-Term Meditators and Matched Nonmeditators"; Siegel, *The Mindful Brain: Reflection and Attunement in the Cultivation of Well-Being*; Pascoe et al., "Mindfulness Mediates the Physiological Markers of Stress: Systematic Review and Meta-Analysis."

139 paying attention to the present moment: Kabat-Zinn, *Full Catastrophe Living: Using the Wisdom of Your Body and Mind to Face Stress, Pain, and Illness.*

140 nine out of ten parents feel judged: Zero to Three, "Judgment."

140 self-compassion benefits our physical and mental health and overall well-being: Biber and Ellis, "The Effect of Self-Compassion on the Self-Regulation of Health Behaviors: A Systematic Review"; Zessin, Dickhäuser, and Garbade, "The Relationship between Self-Compassion and Well-Being: A Meta-Analysis."

140 mindful self-compassion: Neff and Germer, *The Mindful Self-Compassion Workbook: A Proven Way to Accept Yourself, Build Inner Strength, and Thrive*; Neff, *Self-Compassion: The Proven Power of Being Kind to Yourself.*

140 pilot study revealed: Neff and Germer, "A Pilot Study and Randomized Controlled Trial of the Mindful Self-Compassion Program."

141 recognizing that challenges are a shared human experience and that we are not alone: Neff and Germer, *The Mindful Self-Compassion*

Workbook: A Proven Way to Accept Yourself, Build Inner Strength, and Thrive, 35.

141 the self-compassion break: Neff and Germer, *The Mindful Self-Compassion Workbook: A Proven Way to Accept Yourself, Build Inner Strength, and Thrive*, 35.

143 research shows that it can improve your physical and mental health: Neff and Germer, *The Mindful Self-Compassion Workbook: A Proven Way to Accept Yourself, Build Inner Strength, and Thrive*, 35.

143 "breath is a direct pathway to the autonomic nervous system": Dana, *The Polyvagal Theory in Therapy: Engaging the Rhythm of Regulation*, 134.

143 reduce anxiety, stress, and depression: Streeter et al., "Treatment of Major Depressive Disorder with Iyengar Yoga and Coherent Breathing: A Randomized Controlled Dosing Study"; Twal, Wahlquist, and Balasubramanian, "Yogic Breathing When Compared to Attention Control Reduces the Levels of Pro-Inflammatory Biomarkers in Saliva: A Pilot Randomized Controlled Trial"; Ma Xiao et al., "The Effect of Diaphragmatic Breathing on Attention, Negative Affect, and Stress in Healthy Adults."

143 reducing anxiety: Jerath et al., "Self-Regulation of Breathing as a Primary Treatment for Anxiety"; Brown and Gerbarg, *The Healing Power of the Breath: Simple Techniques to Reduce Stress and Anxiety, Enhance Concentration, and Balance Your Emotions.*

143 slow, gentle breathing is calming, helping to reduce anxiety, insomnia, depression, stress, and the effects of trauma: Gerbarg and Brown, "Neurobiology and Neurophysiology of Breath Practices in Psychiatric Care."

144 To achieve an additional sense of calm: Gerbarg and Brown, "Neurobiology and Neurophysiology of Breath Practices in Psychiatric Care"; Dana, *The Polyvagal Theory in Therapy: Engaging the Rhythm of Regulation.*

146 *emotional granularity*: Barrett, *How Emotions Are Made.*

147 Adverse past experiences, such as being mistreated or living in difficult situations: Perry and Szalavitz, *The Boy Who Was Raised as a Dog.*

148 *how we make sense of our life*: Siegel and Hartzell, *Parenting from the Inside Out: How a Deeper Self-Understanding Can Help You Raise Children Who Thrive.*

149 the negativity bias: Rozin and Royzman, "Negativity Bias, Negativity Dominance, and Contagion"; Hanson, *Hardwiring Happiness: The New Brain Science of Contentment, Calm, and Confidence.*

Chapter 6: Making Sense of the Senses

155 "There will never be another you": Eger, "There Will Never Be Another YOU!"

157 *central nervous system*: Stanfield, *Principles of Human Physiology.*

157 *peripheral nervous system*: Stanfield, *Principles of Human Physiology.*

157 constant, bidirectional conversation: Craig, *How Do You Feel? An Interoceptive Moment with Your Neurobiological Self.*

157 The body sends information to the brain: Stanfield, *Principles of Human Physiology.* The pathways from the body to the brain, known as *afferent* neuronal pathways, deliver sensory information such as sound or touch to the brain. The brain then integrates and processes this information, making sense of it, and sends a response via *efferent* neuronal pathways of the vagal nerve back down to the rest of the body.

157 our brain processes that information and sends signals back down to the body, causing our actions: Stanfield, *Principles of Human Physiology.*

158 with some 80 percent of the fibers carrying signals *to* the brain and only 20 percent carrying signals back *from* the brain to the body: Breit et al., "Vagus Nerve as Modulator of the Brain-Gut Axis in Psychiatric and Inflammatory Disorders"; Howland, "Vagus Nerve Stimulation"; Porges, "The Infant's Sixth Sense: Awareness and Regulation of Bodily Processes."

159 respond adaptively to what's required: Ayres, *Sensory Integration and the Child: Understanding Hidden Sensory Challenges.*

159 "You can think of sensations as 'food for the brain'; they provide the knowledge needed to direct the body and mind": Ayres, *Sensory Integration and the Child: Understanding Hidden Sensory Challenges*, 6.

159 "Sensory experiences drive our behavior and contribute to the organization of our thoughts and emotions": Porges, "The Infant's Sixth Sense: Awareness and Regulation of Bodily Processes," 12.

159 inform our reactions to similar experiences in the future: Barrett, *How Emotions Are Made.*

159 "past experiences to construct a hypothesis—the simulation—and [comparing] it to the cacophony arriving from your senses": Barrett, *How Emotions Are Made*, 27.

160 Some children are *over-reactive*, experiencing a sensation more powerfully than most, while others are *under-reactive*: Miller, *Sensational Kids: Hope and Help for Children with Sensory Processing Disorder.*

160 Some have *sensory craving*: Miller, *Sensational Kids: Hope and Help for Children with Sensory Processing Disorder.*

160 *multisensory* experience: Miller, *Sensational Kids: Hope and Help for Children with Sensory Processing Disorder.*

161 our brains are constantly taking in bits of information, comparing it to past experiences, piecing it together, and *integrating it—all* at the same time: Ayres, *Sensory Integration and the Child: Understanding Hidden Sensory Challenges*; Barrett, *How Emotions Are Made.*

161 *sensory integration*: Ayres, *Sensory Integration and the Child: Understanding Hidden Sensory Challenges.*

161 linked children's emotions and behaviors to their complex sensory experiences: Greenspan and Wieder, *Infant and Early Childhood Mental Health: A Comprehensive Developmental Approach to Assessment and Intervention.*

162 it helps us understand how the world *within* our bodies gives rise to our most basic feelings, impacting our emotions and behaviors: Craig, *How Do You Feel? An Interoceptive Moment with Your Neurobiological Self*, 12; Harshaw, "Interoceptive Dysfunction: Toward an Integrated Framework for Understanding Somatic and Affective Disturbance in Depression"; Price and Hooven, "Interoceptive Awareness Skills for Emotion Regulation: Theory and Approach of Mindful Awareness in Body-Oriented Therapy (MABT)"; Mehling et al., "The Multidimensional Assessment of Interoceptive Awareness."

162 Internal sensors send information to the brain: Craig, *How Do You Feel? An Interoceptive Moment with Your Neurobiological Self.*

162 sensations that provide information: Craig, *How Do You Feel? An Interoceptive Moment with Your Neurobiological Self.*

162 *interoceptive awareness*: Craig, *How Do You Feel? An Interoceptive Moment with Your Neurobiological Self.*

162 Interoception: Craig, *How Do You Feel? An Interoceptive Moment with Your Neurobiological Self*; Craig, "How Do You Feel? Interoception: The Sense of the Physiological Condition of the Body"; Barrett, *How Emotions Are Made.*

162 interoceptive awareness to emotions, and emotional regulation: Craig, *How Do You Feel? An Interoceptive Moment with Your Neurobiological Self*; Barrett, *How Emotions Are Made*; Price and Hooven, "Interoceptive Awareness Skills for Emotion Regulation: Theory and Approach of Mindful Awareness in Body-Oriented Therapy (MABT)"; Mahler, *Interoception: The Eighth Sensory System.*

162 interoceptive sensations *generate* our most basic feeling states and moods and, sometimes, what we come to label or identify as emotions: Craig, "How Do You Feel? Interoception: The Sense of the Physiological

Condition of the Body"; Barrett, *How Emotions Are Made*; Price and Hooven, "Interoceptive Awareness Skills for Emotion Regulation: Theory and Approach of Mindful Awareness in Body-Oriented Therapy (MABT)."

162 label those sensations with emotional or other descriptive words: Barrett, *How Emotions Are Made*; Price and Hooven, "Interoceptive Awareness Skills for Emotion Regulation: Theory and Approach of Mindful Awareness in Body-Oriented Therapy (MABT)"; Mahler, *Interoception: The Eighth Sensory System.*

165 emotional granularity: Barrett, *How Emotions Are Made.*

166 sensors in the inner ear: Ayres, *Sensory Integration and the Child: Understanding Hidden Sensory Challenges.*

167 melodic tone of voice: Porges, *The Pocket Guide to the Polyvagal Theory: The Transformative Power of Feeling Safe.*

167 Qualities of the voice help us to judge: Porges, "The Infant's Sixth Sense: Awareness and Regulation of Bodily Processes."

169 cones and rods: Stanfield, *Principles of Human Physiology*; Ayres, *Sensory Integration and the Child: Understanding Hidden Sensory Challenges.*

171 sending information from sensory receptors to the brain: Ayres, *Sensory Integration and the Child: Understanding Hidden Sensory Challenges.*

173 taste receptors: Ayres, *Sensory Integration and the Child: Understanding Hidden Sensory Challenges.*

173 imbued with emotional memories: Greenspan and Wieder, *Infant and Early Childhood Mental Health: A Comprehensive Developmental Approach to Assessment and Intervention*; Barrett, *How Emotions Are Made.*

175 Chemical receptors in the nasal structure: Ayres, *Sensory Integration and the Child: Understanding Hidden Sensory Challenges.*

175 whether something is safe to eat: Ayres, *Sensory Integration and the Child: Understanding Hidden Sensory Challenges.*

177 tells the brain about our body positions: Ayres, *Sensory Integration and the Child: Understanding Hidden Sensory Challenges.*

177 "the brain about when and how the muscles are contracting or stretching, and when and how the joints are bending, extending, or being pulled or compressed": Ayres, *Sensory Integration and the Child: Understanding Hidden Sensory Challenges*, 201.

177 send information to the brain about our body position: Ayres, *Sensory Integration and the Child: Understanding Hidden Sensory Challenges.*

181 our sense of our self: Lopez, "Making Sense of the Body: The Role of Vestibular Signals."

181 Sensors in the inner ear send the brain information: Ayres, *Sensory Integration and the Child: Understanding Hidden Sensory Challenges.*

181 critical system lets you know where your body is in space, whether you are moving or still, how fast you're going, and where you are in relation to gravity: Ayres, *Sensory Integration and the Child: Understanding Hidden Sensory Challenges.*

184 *very foundation of our basic feelings, our bodily sensations*: Price and Hooven, "Interoceptive Awareness Skills for Emotion Regulation: Theory and Approach of Mindful Awareness in Body-Oriented Therapy (MABT)"; Barrett, *How Emotions Are Made.*

184 a strong connection between our ability to detect our body's sensations and our ability to regulate emotions: Price and Hooven, "Interoceptive Awareness Skills for Emotion Regulation: Theory and Approach of Mindful Awareness in Body-Oriented Therapy (MABT)"; Barrett, *How Emotions Are Made.*

185 more experiences of the positive to override the negative memories: Hanson, *Hardwiring Happiness: The New Brain Science of Contentment, Calm, and Confidence.*

Chapter 7: The First Year

189 "Every time a parent looks at that baby and says, 'Oh, you're so wonderful,' that baby just bursts with feeling good": Brazelton, "T. Berry Brazelton Quotes."

191 hormone oxytocin: Carter, "Oxytocin Pathways and the Evolution of Human Behavior."

191 most significant transformation: Brazelton and Sparrow, *Touchpoints: Birth to Three: Your Child's Emotional and Behavioral Development.*

191 fathers who were primary caregivers underwent their own hormonal brain changes: Feldman, "The Adaptive Human Parental Brain: Implications for Children's Social Development"; Abraham et al., "Father's Brain Is Sensitive to Childcare Experiences."

192 finding a shared rhythm: Tronick, *The Neurobehavioral and Social-Emotional Development of Infants and Children.*

192 loving and attuned parents: Shonkoff and Phillips, eds., *From Neurons to Neighborhoods: The Science of Early Childhood Development*; Eshel et al., "Responsive Parenting: Interventions and Outcomes."

192 help babies maintain their body budget: Barrett, *Seven and a Half Lessons about the Brain.*

192 Responsive parents: Shonkoff and Phillips, eds., *From Neurons to Neighborhoods: The Science of Early Childhood Development*; Eshel et al., "Responsive Parenting: Interventions and Outcomes"; Landry et al., "A Responsive Parenting Intervention: The Optimal Timing across Early Childhood for Impacting Maternal Behaviors and Child Outcomes."

192 They tend to do **three things**: Eshel et al., "Responsive Parenting: Interventions and Outcomes"; Landry et al., "A Responsive Parenting Intervention: The Optimal Timing across Early Childhood for Impacting Maternal Behaviors and Child Outcomes."

193 Researchers have found consistently: Eshel et al., "Responsive Parenting: Interventions and Outcomes"; Paul et al., "INSIGHT Responsive Parenting Intervention and Infant Sleep"; Savage et al., "INSIGHT Responsive Parenting Intervention and Infant Feeding Practices: Randomized Clinical Trial"; Hohman et al., "INSIGHT Responsive Parenting Intervention Is Associated with Healthier Patterns of Dietary Exposures in Infants"; Loman and Gunnar, "Early Experience and the Development of Stress Reactivity and Regulation in Children"; Doom and Gunnar, "Stress in Infancy and Early Childhood: Effects on Development"; Gartstein, Hancock, and Iverson, "Positive Affectivity and Fear Trajectories in Infancy: Contributions of Mother-Child Interaction Factors."

193 easier time in preschool: Eshel et al., "Responsive Parenting: Interventions and Outcomes."

193 creating shared meanings: Tronick, *The Neurobehavioral and Social-Emotional Development of Infants and Children*.

193 We are the architects of our baby's experiences and how the baby figures out the world: Tronick and Beeghly, "Infants' Meaning-Making and the Development of Mental Health Problems."

193 the way your baby's brain will make predictions in the future: Barrett, *Seven and a Half Lessons about the Brain*.

194 Growth happens when we try different approaches: Tronick, *The Neurobehavioral and Social-Emotional Development of Infants and Children*.

194 comparing new experiences with old ones: Barrett, *How Emotions Are Made*.

194 updating our child's predictions about us and the world: Barrett, *How Emotions Are Made*.

195 *heart rhythms* sync: Feldman, "Parent–Infant Synchrony: Biological Foundations and Developmental Outcomes"; Feldman et al., "Mother and Infant Coordinate Heart Rhythms through Episodes of Interaction Synchrony."

196 bring their hands near their mouths: Pearson, "Pathways to Positive Parenting: Helping Parents Nurture Healthy Development in the Earliest Months."

197 our life support system: Walker, *Why We Sleep: Unlocking the Power of Sleep and Dreams.*

199 you will tend to have a more prosodic and musical tone of voice: Porges, "Stephen Porges (Polyvagal Perspective and Sound Sensitivity Research)."

199 babies love the sound of a parent's gentle voice: Porges, "Stephen Porges (Polyvagal Perspective and Sound Sensitivity Research)."

202 related to the gastrointestinal system: Johnson, Cocker, and Chang, "Infantile Colic: Recognition and Treatment."

203 an edge in language development, attachment to others, future literacy, and better behavioral and emotional regulation: Mindell and Williamson, "Benefits of a Bedtime Routine in Young Children: Sleep, Development, and Beyond."

204 "the first six months": Task Force on Sudden Infant Death Syndrome.

204 after cycles of 90 to 110 minutes of sleep: National Center for Biotechnology Information, "What Is 'Normal' Sleep?"

205 "the predictable activities that occur in the hour or so before lights out, and before the child falls asleep": Mindell and Williamson, "Benefits of a Bedtime Routine in Young Children: Sleep, Development, and Beyond," 2.

205 Babies and children with predictable bedtime routines sleep better: Mindell et al., "Bedtime Routines for Young Children: A Dose-Dependent Association with Sleep Outcomes."

205 *four basic components*: Mindell and Williamson, "Benefits of a Bedtime Routine in Young Children: Sleep, Development, and Beyond."

208 the brain develops a million connections between neurons: Harvard University, Center on the Developing Child, "Brain Architecture."

209 Follow their lead: Greenspan and Wieder, *Infant and Early Childhood Mental Health: A Comprehensive Developmental Approach to Assessment and Intervention.*

210 powerful building block that sets up the ability to solve problems throughout life: Greenspan and Breslau Lewis, *Building Healthy Minds: The Six Experiences That Create Intelligence and Emotional Growth in Babies and Young Children.*

210 That's why play is called a *neural or brain exercise*: Porges, *The Pocket Guide to the Polyvagal Theory: The Transformative Power of Feeling Safe.*

Chapter 8: Tantrums Throw Toddlers

213 "Calmly support children through their frustration, disappointment and even failure, so that we normalize these difficult but healthy life experiences": Lansbury, *Elevating Child Care: A Guide to Respectful Parenting.*

217 *begins* to emerge in toddlerhood and continues all the way through the mid-twenties: Harvard University, Center on the Developing Child, "The Science of Adult Capabilities."

217 statistical learning: Barrett, *How Emotions Are Made*, 94.

217 still learning to make accurate predictions: Barrett, *How Emotions Are Made.*

218 The expectation gap: Zero to Three, "Parent Survey Reveals Expectation Gap for Parents of Young Children."

219 high rates of suspensions and expulsions of little ones from preschool in the United States: Malik, "New Data Reveal 250 Preschoolers Are Suspended or Expelled Every Day."

224 regulating or tempering according to measure or proportion: *Merriam-Webster*, online, s.v. "modulation."

225 far more likely to be targeted for punishment for his behavior problems: United States Department of Education Office for Civil Rights, "Data Snapshot: Early Childhood Education"; Malik, "New Data Reveal 250 Preschoolers Are Suspended or Expelled Every Day."

226 facial expressions do not universally match up with emotion states: Barrett, *How Emotions Are Made*; Fox et al., *The Nature of Emotion: Fundamental Questions.*

227 built primarily through relationships: Shonkoff and Phillips, eds., *From Neurons to Neighborhoods: The Science of Early Childhood Development.*

227 helping children take note of their bodies' sensations: Barrett, *How Emotions Are Made*; Mahler, *Interoception: The Eighth Sensory System.*

227 *using words and concepts to understand our experiences and share them with others*: Greenspan and Wieder, *Infant and Early Childhood Mental Health: A Comprehensive Developmental Approach to Assessment and Intervention*; Barrett, *How Emotions Are Made.*

229 metaphorically reverberate or vibrate with a similar energy or emotional level as the child: Stanford Medicine, The Center for Compassion and Altruism Research and Education, "Emotion Resonance."

229 *tether to the grounded state of the nervous system*: Dana, "Reaching Out in Nervous Times: Polyvagal Theory Encounters Teletherapy."

230 "it's not what you say, it's how you say it": Porges, "The Neurophysiology of Safety and How to Feel Safe."

231 begin to use concepts: Barrett, *How Emotions Are Made.*

235 build bridges between our ideas and those of others: Greenspan and Wieder, *Infant and Early Childhood Mental Health: A Comprehensive Developmental Approach to Assessment and Intervention.*

Chapter 9: Elementary School-Age Kids

245 "Emotional literacy helps your emotions work for you and not against you": Steiner, in "What Is Emotional Literacy?"

248 recently recommended that pediatricians prescribe play: Yogman et al., "The Power of Play: A Pediatric Role in Enhancing Development in Young Children."

249 "research demonstrates that developmentally appropriate play with parents and peers is a singular opportunity to promote the social-emotional, cognitive, language, and self-regulation skills that build executive function and a pro-social brain": Yogman et al., "The Power of Play: A Pediatric Role in Enhancing Development in Young Children," 1.

249 high-quality school recess contributes significantly to children's executive functioning: Massey and Geldhof, "High Quality Recess Contributes to the Executive Function, Emotional Self-Control, Resilience, and Positive Classroom Behavior in Elementary School Children."

250 *symbolic play*: Greenspan and Wieder, *Infant and Early Childhood Mental Health: A Comprehensive Developmental Approach to Assessment and Intervention.*

251 play reveals the issues and concerns on a child's mind: Greenspan and Wieder, *Infant and Early Childhood Mental Health: A Comprehensive Developmental Approach to Assessment and Intervention.*

251 stimulating cognitive development in preschoolers: Ngan Kuen Lai et al., "The Impact of Play on Child Development—A Literature Review."

254 *Follow their lead*: Greenspan and Wieder, *Infant and Early Childhood Mental Health: A Comprehensive Developmental Approach to Assessment and Intervention.*

256 *expanding* the play: Greenspan and Wieder, *Infant and Early Childhood Mental Health: A Comprehensive Developmental Approach to Assessment and Intervention*; Greenspan, *Infancy and Early Childhood: The Practice of Clinical Assessment and Intervention with Emotional and Developmental Challenges.*

257 rivalry, jealousy, anger, and sadness: Greenspan and Wieder, *Infant and Early Childhood Mental Health: A Comprehensive Developmental Approach to Assessment and Intervention.*

259 *neural, or brain, exercise*: Porges, *The Pocket Guide to the Polyvagal Theory: The Transformative Power of Feeling Safe.*

264 befriend their nervous systems: Dana, *The Polyvagal Theory in Therapy: Engaging the Rhythm of Regulation.*

264 feeling of sensations inside the body: Craig, *How Do You Feel? An Interoceptive Moment with Your Neurobiological Self*; Porges, "The Infant's Sixth Sense: Awareness and Regulation of Bodily Processes."

264 "an emotion is your brain's creation of what your *bodily sensations mean*, in relation to what's going on around you in the world": Barrett, *How Emotions Are Made*, 30.

264 leads to improved emotional regulation: Price and Hooven, "Interoceptive Awareness Skills for Emotion Regulation: Theory and Approach of Mindful Awareness in Body-Oriented Therapy (MABT)"; Craig, *How Do You Feel? An Interoceptive Moment with Your Neurobiological Self*; Barrett, *How Emotions Are Made.*

264 "How do you feel?": Craig, *How Do You Feel? An Interoceptive Moment with Your Neurobiological Self.*

265 *helping them tune in to their body's sensations and make sense of them with self-compassion*: Price and Hooven, "Interoceptive Awareness Skills for Emotion Regulation: Theory and Approach of Mindful Awareness in Body-Oriented Therapy (MABT)"; Mahler, *Interoception: The Eighth Sensory System.*

265 redefining how we understand our emotions and actions: Barrett, *How Emotions Are Made.*

266 connecting an experience or physical sensation to a word: Barrett, *How Emotions Are Made.*

267 *sensation-to-feeling-to-emotion process*: Price and Hooven, "Interoceptive Awareness Skills for Emotion Regulation: Theory and Approach of Mindful Awareness in Body-Oriented Therapy (MABT)."

267 *emotional intelligence*: Salovey and Mayer, "Emotional Intelligence."

268 *recast sensations*: Mahler, *Interoception: The Eighth Sensory System.*

269 *emotional granularity*: Barrett, *How Emotions Are Made.*

271 Body's Signals: Adapted from Delahooke, *Beyond Behaviors: Using Brain Science and Compassion to Understand and Solve Children's Behavioral Challenges.*

Chapter 10: Flourishing

279 "My mission in life is not merely to survive, but to thrive; and to do so with some passion, some compassion, some humor, and some style": Angelou, "Maya Angelou: In Her Own Words."

283 Danish word for coziness, *hygge*: Wiking, *The Little Book of Hygge: Danish Secrets to Happy Living.*

283 "Time spent with others creates an atmosphere that is warm, relaxed, friendly, down-to-earth, close, comfortable, snug, and welcoming. In many ways, it is like a good hug, but without the physical contact. It is in this situation that you can be completely relaxed and yourself": Wiking, *The Little Book of Hygge: Danish Secrets to Happy Living*, 39.

283 *gezellig*: Phillips, "Move Over, Hygge, Gezellig Is the Trendy Danish Lifestyle Philosophy to Try."

283 evolved through four major waves: Al-Taher, "The 5 Founding Fathers and a History of Positive Psychology."

284 Skinner, who believed that we could modify all human behaviors through the use of rewards and punishments, or consequences: Al-Taher, "The 5 Founding Fathers and a History of Positive Psychology."

284 positive psychology movement: Al-Taher, "The 5 Founding Fathers and a History of Positive Psychology."

284 The term became popular: Fishman, "Positive Psychology: The Benefits of Living Positively."

284 how they feel *at the moment they take the survey*: Seligman, *Flourish: A Visionary New Understanding of Happiness and Well-Being.*

285 PERMA model: Seligman, *Flourish: A Visionary New Understanding of Happiness and Well-Being.*

285 So does *engagement*: Seligman, *Flourish: A Visionary New Understanding of Happiness and Well-Being.*

285 this state is referred to as *flow*: Ackerman, "Flourishing in Positive Psychology: Definition + 8 Practical Tips."

285 being "in the zone," when you're so engaged in an activity that you can lose track of time: Seligman, *Flourish: A Visionary New Understanding of Happiness and Well-Being.*

285 to feel a sense of *accomplishment* and that what we do in our life matters: Seligman, *Flourish: A Visionary New Understanding of Happiness and Well-Being.*

BIBLIOGRAPHY

Abraham, Eyal, Talma Hendler, Irit Shapira-Lichter, Yaniv Kanat-Maymon, Orna Zagoory-Sharon, and Ruth Feldman. "Father's Brain Is Sensitive to Childcare Experiences." *Proceedings of the National Academy of Sciences* 111, no. 27 (May 2014): 9792–97. https://doi.org/10.1073/pnas.1402569111.

Ackerman, Courtney E. "Flourishing in Positive Psychology: Definition + 8 Practical Tips." PositivePsychology.com, January 25, 2021. https://positive psychology.com/flourishing/.

Al-Taher, Reham. "The 5 Founding Fathers and a History of Positive Psychology." PositivePsychology.com, May 7, 2021. https://positivepsychology.com /founding-fathers/.

Angelou, Maya. "Maya Angelou: In Her Own Words." BBC News, May 28, 2014. https://www.bbc.com/news/world-us-canada-27610770.

APA Dictionary of Psychology. "Frustration Tolerance." American Psychological Association. Accessed May 13, 2021. https://dictionary.apa.org/frustration -tolerance.

Artuch-Garde, Raquel, Maria del Carmen Gonzalez-Torres, Jesus de la Fuente, M. Mariano Vera, Maria Fernandez-Cabezas, and Mireia Lopez-Garcia. "Relationship between Resilience and Self-Regulation: A Study of Spanish Youth at Risk of Social Exclusion." *Frontiers in Psychology* (April 2017). https://dx.doi.org/10.3389%2Ffpsyg.2017.00612.

Associated Press. "Fauci Says Pandemic Exposed 'Undeniable Effects of Racism.'" *Los Angeles Times*, May 16, 2021. https://www.latimes.com/world-nation /story/2021-05-16/fauci-covid19-pandemic-racism.

Ayres, A. Jean. *Sensory Integration and the Child: Understanding Hidden Sensory Challenges*. 25th anniversary ed. Los Angeles: Western Psychological Services, 2005.

Baer, Ruth A., Emily L. B. Lykins, and Jessica R. Peters. "Mindfulness and Self-Compassion as Predictors of Psychological Wellbeing in Long-Term

Meditators and Matched Nonmeditators." *Journal of Positive Psychology* 7, no. 3 (2012): 230–38. https://doi.org/10.1080/17439760.2012.674548.

Barrett, Lisa Feldman. "Bloom Where You Are Rooted: What Neuroscience Can Teach Us about Harnessing Passion and Productivity." Interview by Lisa Cypers Kamen. *Harvesting Happiness* podcast, November 25, 2020. https://soundcloud.com/lisa-cypers-kamen/bloom-where-you-are-rooted-what-neuroscience-can-teach-us-about-harnessing-passion-and-productivity.

Barrett, Lisa Feldman. *How Emotions Are Made: The Secret Life of the Brain.* New York: Houghton Mifflin Harcourt, 2017.

Barrett, Lisa Feldman. "Neuroscientist Reveals Your Brain Is Just 'Guessing' & Doesn't Know Anything." Interview by Tom Bilyeu. *Impact Theory* podcast, November 12, 2020. https://podcasts.google.com/feed/aHR0cHM6Ly9pbXBhY3R0aGVvcnkubGlic3luLmNvbS9yc3M/episode/ZGQzODcwNzAtZDdhNS00ZjlkLWFkYTEtNDJhMjY0ZGRhY2Fh.

Barrett, Lisa Feldman. *Seven and a Half Lessons about the Brain.* New York: Houghton Mifflin Harcourt, 2020.

Barrett, Lisa Feldman. "The Theory of Constructed Emotion: An Active Inference Account of Interoception and Categorization." *Social Cognitive and Affective Neuroscience* 12, no. 11 (January 2017): 1–23. https://doi.org/10.1093/scan/nsw154.

Benjamin, Courtney L., Kelly A. O'Neil, Sarah A. Crawley, Rinad S. Beidas, Meredith Coles, and Philip C. Kendall. "Patterns and Predictors of Subjective Units of Distress in Anxious Youth." *Behavioural and Cognitive Psychotherapy* 38, no. 4 (July 2010): 497–504. https://doi.org/10.1017/s1352465810000287.

Berntson, Gary G., and John T. Cacioppo. "Heart Rate Variability: Stress and Psychiatric Conditions." In *Dynamic Electrocardiography*, edited by Marek Malik and A. John Camm, 57–64. Oxford, UK: Blackwell Publishing, 2004.

Berntson, Gary G., John T. Cacioppo, and Karen S. Quigley. "Autonomic Determinism: The Modes of Autonomic Control, the Doctrine of Autonomic Space, and the Laws of Autonomic Constraint." *Psychological Review* 98, no. 4 (1991): 459–87, https://doi.org/10.1037/0033-295X.98.4.459.

Biber, David D., and Rebecca Ellis. "The Effect of Self-Compassion on the Self-Regulation of Health Behaviors: A Systematic Review." *Journal of Health Psychology* 24, no. 14 (June 2017): 2060–71. https://doi.org/10.1177/1359105317713361.

Brazelton, T. Berry. "T. Berry Brazelton Quotes." Quote Fancy. Accessed May 18, 2021. https://quotefancy.com/t-berry-brazelton-quotes.

Brazelton, T. Berry, and Joshua D. Sparrow. *Touchpoints: Birth to Three: Your Child's Emotional and Behavioral Development*. 2nd ed. Cambridge, MA: Perseus Group, 2006.

Breit, S., A. Kupferberg, G. Rogler, and G. Hasler. "Vagus Nerve as Modulator of the Brain-Gut Axis in Psychiatric and Inflammatory Disorders." *Frontiers in Psychiatry* 9, no. 44 (2018). doi:10.3389/fpsyt.2018.00044.

Brooks, A. W. "Get Excited: Reappraising Pre-Performance Anxiety as Excitement." *Journal of Experimental Psychology: General* 143, no. 3 (2014): 1144–58. https://doi.org/10.1037/a0035325.

Brown, Richard, and Patricia Gerbarg. *The Healing Power of the Breath: Simple Techniques to Reduce Stress and Anxiety, Enhance Concentration, and Balance Your Emotions.* Boston: Shambhala, 2012.

Bush, Nicole R., Karen Jones-Mason, Michael Coccia, Zoe Caron, Abbey Alkon, Melanie Thomas, Elissa Epel, et al. "Effects of Pre- and Postnatal Maternal Stress on Infant Temperament and Autonomic Nervous System Reactivity and Regulation in a Diverse, Low-Income Population." *Development and Psychopathology* 29, no. 5 (December 2017): 1553–71. https://doi.org/10.1017/s0954579417001237.

Cacioppo, John T., and Stephanie Cacioppo. "The Growing Problem of Loneliness." *The Lancet* 391, no. 10119 (February 2018): 426. https://doi.org/10.1016/S0140-6736(18)30142-9.

Carter, C. Sue. "Oxytocin Pathways and the Evolution of Human Behavior." *Annual Review of Psychology* 65, no. 1 (September 2013): 17–39. https://doi.org/10.1146/annurev-psych-010213-115110.

Christensen, Jacquelyn S., Heather Wild, Erin S. Kenzie, Wayne Wakeland, Deborah Budding, and Connie Lillas. "Diverse Autonomic Nervous System Stress Response Patterns in Childhood Sensory Modulation." *Frontiers in Integrative Neuroscience* (February 2020). https://doi.org/10.3389/fnint.2020.00006.

Cigna. "Cigna U.S. Loneliness Index." May 2018. Accessed May 18, 2021. https://www.cigna.com/assets/docs/newsroom/loneliness-survey-2018-full-report.pdf.

Clark, Andy. "Whatever Next? Predictive Brains, Situated Agents, and the Future of Cognitive Science." *Behavioral and Brain Sciences* 36, no. 3 (June 2013): 181–204. https://doi.org/10.1017/S0140525X12000477.

Cleveland Clinic. "Stress." Last reviewed January 28, 2021. Accessed May 13, 2021. https://my.clevelandclinic.org/health/articles/11874-stress.

Craig, A. D. *How Do You Feel? An Interoceptive Moment with Your Neurobiological Self.* Princeton, NJ: Princeton University Press, 2014.

Craig, A. D. "How Do You Feel? Interoception: The Sense of the Physiological Condition of the Body." *Nature Reviews Neuroscience* 3 (August 2002): 655–66. https://doi.org/10.1038/nrn894.

Crosswell, Alexandra D., Michael Coccia, and Elissa Epel. "Mind Wandering and Stress: When You Don't Like the Present Moment." *Emotion* 20, no. 3 (April 2020): 403–12. https://doi.org/10.1037/emo0000548.

Dana, Deb. *The Polyvagal Theory in Therapy: Engaging the Rhythm of Regulation.* New York: W. W. Norton, 2018.

Dana, Deb. "Reaching Out in Nervous Times: Polyvagal Theory Encounters Teletherapy." *Psychotherapy Networker*, November/December 2020. https://www.psychotherapynetworker.org/magazine/article/2507/reaching-out-in-nervous-times.

DeGangi, Georgia A., Janet A. Dipietro, Stanley I. Greenspan, and Stephen Porges. "Psychophysiological Characteristics of the Regulatory Disordered Infant." *Infant Behavior and Development* 14, no. 1 (1991): 37–50. https://doi.org/10.1016/0163-6383(91)90053-U.

Delahooke, Mona. *Beyond Behaviors: Using Brain Science and Compassion to Understand and Solve Children's Behavioral Challenges.* Eau Claire, WI: PESI Publishing and Media, 2019.

Diamond, Adele. "Executive Functions." *Annual Review of Psychology* 64 (2013): 135–68. https://doi.org/10.1146/annurev-psych-113011-143750.

Di Pellegrino, G., L. Fadiga, L. Fogassi, V. Gallese, and G. Rizzolatti. "Understanding Motor Events: A Neurophysiological Study." *Experimental Brain Research* 91 (October 1992): 176–80. https://doi.org/10.1007/BF00230027.

Doom, J. R., and M. R. Gunnar. "Stress in Infancy and Early Childhood: Effects on Development." In *International Encyclopedia of the Social & Behavioral Sciences.* 2nd ed. Edited by James D. Wright, 577–82. Oxford: Elsevier, 2015.

Duckworth, Angela. *Grit: The Power of Passion and Perseverance.* New York: Simon & Schuster, 2016.

Duckworth, Angela, and Stephanie M. Carlson. "Self-Regulation and School Success." In *Self-Regulation and Autonomy: Social and Developmental Dimensions of Human Conduct.* Edited by Bryan W. Sokol, Frederick M. E. Grouzet, and Ulrich Müller, 208–30. New York: Cambridge University Press, 2013.

Ebisch, Sjoerd J., Tiziana Aureli, Daniela Bafunno, Daniela Cardone, Gian Luca Romani, and Arcangelo Merla. "Mother and Child in Synchrony: Thermal Facial Imprints of Autonomic Contagion." *Biological Psychology* 89, no. 1 (January 2012): 123–29. https://doi.org/10.1016/j.biopsycho.2011.09.018.

Eger, Edith. "There Will Never Be Another YOU!" Video. Posted September 29, 2019. https://www.facebook.com/GenWNow/videos/807562599641758.

"Embrace2." Empatica. Accessed May 25, 2021. https://www.empatica.com/embrace2/.

Epel, Elissa, Jennifer Daubenmier, Judith T. Moskowitz, Susan Folkman, and Elizabeth Blackburn. "Can Meditation Slow Rate of Cellular Aging? Cognitive Stress, Mindfulness, and Telomeres." *Annals of the New York Academy of Sciences* 1172, no. 1 (August 2009): 34–53. https://dx.doi.org/10.1111%2Fj.1749-6632.2009.04414.x.

Epel, Elissa, Eli Puterman, Jue Lin, Elizabeth Blackburn, Alanie Lazaro, and Wendy Berry Mendes. "Wandering Minds and Aging Cells." *Clinical Psychological Science* 1, no. 1 (November 2012): 75–83. https://doi.org/10.1177/2167702612460234.

Eshel, Neir, Bernadette Daelmans, Meena Cabral de Mello, and Jose Martines. "Responsive Parenting: Interventions and Outcomes." *Bulletin of the World Health Organization* 84 (2006): 992–99. https://doi.org/10.2471/blt.06.030163.

Feldman, Ruth. "The Adaptive Human Parental Brain: Implications for Children's Social Development." *Trends in Neurosciences* 38, no. 6 (June 2015): 387–99. https://doi.org/10.1016/j.tins.2015.04.004.

Feldman, Ruth. "Infant–Mother and Infant–Father Synchrony: The Coregulation of Positive Arousal." *Infant Mental Health Journal* 24, no. 1 (January/February 2003): 1–23. https://doi.org/10.1002/imhj.10041.

Feldman, Ruth. "Parent–Infant Synchrony: Biological Foundations and Developmental Outcomes." *Current Directions in Psychological Science* 16, no. 6 (December 2007): 340–45. https://doi.org/10.1111/j.1467-8721.2007.00532.x.

Feldman, Ruth, Romi Magori-Cohen, Giora Galili, Magi Singer, and Yoram Louzoun. "Mother and Infant Coordinate Heart Rhythms through Episodes of Interaction Synchrony." *Infant Behavior and Development* 34, no. 4 (December 2011): 569–77. https://doi.org/10.1016/j.infbeh.2011.06.008.

Fishman, Joanna. "Positive Psychology: The Benefits of Living Positively." PsychCentral, March 11, 2013. https://psychcentral.com/blog/positive-psychology-the-benefits-of-living-positively#1.

Fogel, Alan. *Developing through Relationships: Origins of Communication, Self, and Culture*. Chicago: University of Chicago Press, 1993.

Fox, Andrew S., Regina C. Lapate, Alexander J. Shackman, and Richard J. Davidson, eds. *The Nature of Emotion: Fundamental Questions*. New York: Oxford University Press, 2018.

Fraley, Chris R., Glenn I. Roisman, and John D. Haltigan. "The Legacy of Early Experiences in Development: Formalizing Alternative Models of How Early Experiences Are Carried Forward over Time." *Developmental Psychology* 49, no. 1 (January 2013): 109–26. https://doi.org/10.1037/a0027852.

Garber, Benjamin D. "For the Love of Fluffy: Respecting, Protecting, and Empowering Transitional Objects in the Context of High-Conflict Divorce." *Journal of Divorce & Remarriage* 60, no. 7 (2019): 552–65. https://doi.org/10.1080/10502556.2019.1586370.

Gartstein, Maria A., Gregory R. Hancock, and Sydney L. Iverson. "Positive Affectivity and Fear Trajectories in Infancy: Contributions of Mother-Child Interaction Factors." *Child Development* 89, no. 5 (September 2018): 1519–34. https://doi.org/10.1111/cdev.12843.

Gerbarg, Patricia, and Richard Brown. "Neurobiology and Neurophysiology of Breath Practices in Psychiatric Care." *Psychiatric Times* 33, no. 11 (November 2016). https://www.psychiatrictimes.com/view/neurobiology-and-neurophysiology-breath-practices-psychiatric-care.

Gianino, A., and E. Z. Tronick. "The Mutual Regulation Model: The Infant's Self and Interactive Regulation and Coping and Defensive Capacities." In *Stress and Coping across Development*. Edited by Tiffany M. Field, Philip McCabe, and Neil Schneiderman, 47–68. Hillsdale, NJ: Erlbaum, 1988.

Gillespie, Linda. "It Takes Two: The Role of Co-Regulation in Building Self-Regulation Skills." Zero to Three. https://www.zerotothree.org/resources/1777-it-takes-two-the-role-of-co-regulation-in-building-self-regulation-skills.

Goldstein, Andrea N., and Matthew P. Walker. "The Role of Sleep in Emotional Brain Function." *Annual Review of Clinical Psychology* 10 (March 2014): 679–708. https://dx.doi.org/10.1146%2Fannurev-clinpsy-032813-153716.

Gray, Sarah A. O., Christopher W. Jones, Katherine P. Theall, Erin Glackin, and Stacy S. Drury. "Thinking across Generations: Unique Contributions of Maternal Early Life and Prenatal Stress to Infant Physiology." *Journal of the American Academy of Child and Adolescent Psychiatry* 56, no. 11 (November 2017): 922–29. https://doi.org/10.1016/j.jaac.2017.09.001.

Greene, Ross. *The Explosive Child: A New Approach for Understanding and Parenting Easily Frustrated, Chronically Inflexible Children.* New York: HarperCollins, 1998.

Greenspan, Stanley. "The Greenspan Floortime Approach: Interaction." The Greenspan Floortime Approach. Accessed May 13, 2021. https://www.stanleygreenspan.com/tags/interaction.

Greenspan, Stanley. *Infancy and Early Childhood: The Practice of Clinical Assessment and Intervention with Emotional and Developmental Challenges.* Madison, CT: International Universities Press, 1992.

Greenspan, Stanley, and Nancy Breslau Lewis. *Building Healthy Minds: The Six Experiences That Create Intelligence and Emotional Growth in Babies and Young Children.* New York: Da Capo Press, 1999.

Greenspan, Stanley, and Serena Wieder. *Infant and Early Childhood Mental Health: A Comprehensive Developmental Approach to Assessment and Intervention.* Washington, DC: American Psychiatric Publishing, 2006.

Hanson, Rick. *Buddha's Brain: The Practical Neuroscience of Happiness, Love, and Wisdom.* Oakland, CA: New Harbinger Publications, 2009.

Hanson, Rick. *Hardwiring Happiness: The New Brain Science of Contentment, Calm, and Confidence.* New York: Random House, 2013.

Harshaw, Christopher. "Interoceptive Dysfunction: Toward an Integrated Framework for Understanding Somatic and Affective Disturbance in

Depression." *Psychological Bulletin* 141, no. 2 (March 2015): 311–63. https://doi.org/10.1037/a0038101.

Harvard University, Center on the Developing Child. "Brain Architecture." Accessed May 18, 2021. https://developingchild.harvard.edu/science/key-concepts/brain-architecture/#neuron-footnote.

Harvard University, Center on the Developing Child. "Resilience." Accessed May 17, 2021. https://developingchild.harvard.edu/science/key-concepts/resilience/.

Harvard University, Center on the Developing Child. "The Science of Adult Capabilities." Accessed May 17, 2021. https://developingchild.harvard.edu/science/deep-dives/adult-capabilities/.

Harvard University, Center on the Developing Child. "Serve and Return." Accessed May 17, 2021. https://developingchild.harvard.edu/science/key-concepts/serve-and-return/.

Hatfield, Elaine, Lisamarie Bensman, Paul Thornton, and Richard Rapson. "New Perspectives on Emotional Contagion: A Review of Classic and Recent Research on Facial Mimicry and Contagion." *Interpersona: An International Journal on Personal Relationships* 8, no. 2 (December 2014): 159–79. https://doi.org/10.5964/ijpr.v8i2.162.

Hohman, Emily E., Ian M. Paul, Leann L. Birch, and Jennifer S. Savage. "INSIGHT Responsive Parenting Intervention Is Associated with Healthier Patterns of Dietary Exposures in Infants." *Obesity* 25, no. 1 (2017): 185–91. https://dx.doi.org/10.1002%2Foby.21705.

Hong-feng Gu, Chao-ke Tang, and Yong-zong Yang. "Psychological Stress, Immune Response, and Atherosclerosis." *Atherosclerosis* 223, no. 1 (July 2012): 69–77. https://doi.org/10.1016/j.atherosclerosis.2012.01.021.

Howland, Robert. "Vagus Nerve Stimulation." *Current Behavioral Neuroscience Reports* 1, no. 2 (June 2014): 64–73. https://doi.org/10.1007/s40473-014-0010-5.

Iacoboni, Marco. *Mirroring People: The Science of Empathy and How We Connect with Others*. New York: Picador, 2008.

Immordino-Yang, Mary Helen, Linda Darling-Hammond, and Christina R. Krone. "Nurturing Nature: How Brain Development Is Inherently Social and Emotional, and What This Means for Education." *Educational Psychologist* 54, no. 3 (2019): 185–204. https://doi.org/10.1080/00461520.2019.1633924.

Jerath, Ravinder, Molly W. Crawford, Vernon A. Barnes, and Kyler Harden. "Self-Regulation of Breathing as a Primary Treatment for Anxiety." *Applied Psychophysiology and Biofeedback* 40, no. 2 (2015): 107–15. https://doi.org/10.1007/s10484-015-9279-8.

Johnson, Jeremy D., Katherine Cocker, and Elisabeth Chang. "Infantile Colic:

Recognition and Treatment." *American Family Physician* 92, no. 7 (October 2015): 577–82. https://www.aafp.org/afp/2015/1001/p577.html.

Kabat-Zinn, Jon. *Full Catastrophe Living: Using the Wisdom of Your Body and Mind to Face Stress, Pain, and Illness.* New York: Random House, 1990.

Kendo Notes Blog. "'Soft Eyes,' a Way of Seeing and Being—Quotes and Resources." Posted December 21, 2018. https://kendonotes.wordpress.com/2018/12/21/quotes-on-soft-eyes-a-way-to-see/.

Keysers, Christian, and Valeria Gazzola. "Hebbian Learning and Predictive Mirror Neurons for Actions, Sensations, and Emotions." *Philosophical Transactions of the Royal Society of London, Series B, Biological Sciences* 369, no. 1644 (2014). https://doi.org/10.1098/rstb.2013.0175.

Koulivand, Peir Hossein, Maryam Khaleghi Ghadiri, and Ali Gorji. "Lavender and the Nervous System." *Evidence-Based Complementary and Alternative Medicine* (2013). https://dx.doi.org/10.1155%2F2013%2F681304.

Landry, Susan H., Karen E. Smith, Paul R. Swank, and Cathy Guttentag. "A Responsive Parenting Intervention: The Optimal Timing across Early Childhood for Impacting Maternal Behaviors and Child Outcomes." *Developmental Psychology* 44, no. 5 (December 2008): 1335–53. https://dx.doi.org/10.1037%2Fa0013030.

Lansbury, Janet. *Elevating Child Care: A Guide to Respectful Parenting.* Los Angeles: JLML Press, 2014.

Levine, Peter. *Healing Trauma: A Pioneering Program for Restoring the Wisdom of Your Body.* Aurora, CA: Sounds True, 2008.

Lillas, Connie. "Handouts." Neurorelational Framework Institute. October 18, 2021. https://nrfr2r.com/for-families/.

Lillas, Connie. "NRF Foundations Manual" (forthcoming). Neurorelational Framework Institute. October 18, 2021. https://nrfr2r.com/nrf-manuals/.

Lillas, Connie, and Janiece Turnbull. *Infant/Child Mental Health, Early Intervention, and Relationship-Based Therapies: A Neurorelational Framework for Interdisciplinary Practice.* New York: W. W. Norton, 2009.

Loman, Michelle M., and Megan R. Gunnar. "Early Experience and the Development of Stress Reactivity and Regulation in Children." *Neuroscience and Biobehavioral Reviews* 34, no. 6 (2010): 867–76. https://doi.org/10.1016/j.neubiorev.2009.05.007.

Lopez, Christophe. "Making Sense of the Body: The Role of Vestibular Signals." *Multisensory Research* 28 no. 5–6 (July 2015): 525–57. https://doi.org/10.1163/22134808-00002490.

Luthar, Suniya S., and Lucia Ciciolla. "Who Mothers Mommy? Factors That Contribute to Mothers' Well-Being." *Developmental Psychology* 51, no. 12 (2015): 1812–23. https://doi.org/10.1037/dev0000051.

Ma Xiao, Yue Zi-Qi, Gong Zhu-Qing, Zhang Hong, Duan Nai-Yue, Shi Yu-Tong, Wei Gao-Xia, and Li You-Fa. "The Effect of Diaphragmatic Breathing on

Attention, Negative Affect, and Stress in Healthy Adults." *Frontiers in Psychology* 8, no. 874 (June 2017). https://dx.doi.org/10.3389%2Ffpsyg.2017.00874.

Mahler, Kelly. *Interoception: The Eighth Sensory System*. Shawnee Mission, KS: AAPC Publishing, 2015.

Malik, Rasheed. "New Data Reveal 250 Preschoolers Are Suspended or Expelled Every Day." Center for American Progress, November 6, 2017. https://www.americanprogress.org/issues/early-childhood/news/2017/11/06/442280/new-data-reveal-250-preschoolers-suspended-expelled-every-day/.

Massey, William, and John Geldhof. "High Quality Recess Contributes to the Executive Function, Emotional Self-Control, Resilience, and Positive Classroom Behavior in Elementary School Children." Study by Oregon State University, College of Public Health and Human Sciences, October 2019. https://www.playworks.org/wp-content/uploads/2019/12/Recess-Outcomes-Study-2019-One-Pager-Only-v3.pdf.

Mayo Clinic. "Nutrition and Healthy Eating." Accessed May 17, 2021. https://www.mayoclinic.org/healthy-lifestyle/nutrition-and-healthy-eating/in-depth/water/art-20044256.

McClelland, Megan M., Claire Cameron Ponitz, Emily E. Messersmith, and Shauna Tominey. "Self-Regulation: The Integration of Cognition and Emotion." In *The Handbook of Life-Span Development, Volume 1: Cognition, Biology, and Methods*. Edited by Richard Lerner and Willis Overton, 509–53. Hoboken, NJ: Wiley, 2010.

McEwen, Bruce. "Protective and Damaging Effects of Stress Mediators." *New England Journal of Medicine* 338, no. 3 (January 1998): 171–79. https://doi.org/10.1056/nejm199801153380307.

McEwen, Bruce. "Protective and Damaging Effects of Stress Mediators: Central Role of the Brain." *Dialogues in Clinical Neuroscience* 8, no. 4 (December 2006): 367–81. https://doi.org/10.31887/DCNS.2006.8.4/bmcewen.

McEwen, Bruce, and Elizabeth Norton Lasley. *The End of Stress as We Know It*. Washington, DC: Joseph Henry Press, 2002.

Mehling, Wolf E., Cynthia Price, Jennifer J. Daubenmier, Mike Acree, Elizabeth Bartmess, and Anita Stewart. "The Multidimensional Assessment of Interoceptive Awareness." *PLoS ONE* 7, no. 11 (2012). https://doi.org/10.1371/journal.pone.0048230.

Merriam-Webster, online, s.v. "modulation." Accessed May 18, 2021. https://www.merriam-webster.com/dictionary/modulation.

Miller, Lucy Jane. *Sensational Kids: Hope and Help for Children with Sensory Processing Disorder*. New York: Penguin Books, 2007.

Mindell, Jodi A., Albert M. Li, Avi Sadeh, Robert Kwon, and Daniel Y. T. Goh. "Bedtime Routines for Young Children: A Dose-Dependent Association with Sleep Outcomes." *Sleep* 38, no. 5 (May 2015): 717–22. https://doi.org/10.5665/sleep.4662.

Mindell, Jodi A., and Ariel A. Williamson. "Benefits of a Bedtime Routine in Young Children: Sleep, Development, and Beyond." *Sleep Medicine Reviews* 40, no. 93 (August 2018): 93–108. https://doi.org/10.1016/j.smrv.2017.10.007.

Montroy, J. J., R. P. Bowles, L. E. Skibbe, M. M. McClelland, and F. J. Morrison. "The Development of Self-Regulation across Early Childhood." *Developmental Psychology* 52, no. 11 (2016): 1744–62. https://doi.org/10.1037/dev0000159.

National Academies of Sciences, Engineering, and Medicine. *Vibrant and Healthy Kids: Aligning Science, Practice, and Policy to Advance Health Equity*. Washington, DC: The National Academies Press, 2019. https://doi.org/10.17226/25466.

National Center for Biotechnology Information. "What Is 'Normal' Sleep?" Via InformedHealth.org and Institute for Quality and Efficiency in Health Care. Accessed May 18, 2021. https://www.ncbi.nlm.nih.gov/books/NBK279322/.

Neff, Kristin. *Self-Compassion: The Proven Power of Being Kind to Yourself.* New York: HarperCollins, 2011.

Neff, Kristin, and Christopher Germer. *The Mindful Self-Compassion Workbook: A Proven Way to Accept Yourself, Build Inner Strength, and Thrive.* New York: Guilford Press, 2018.

Neff, Kristin, and Christopher Germer. "A Pilot Study and Randomized Controlled Trial of the Mindful Self-Compassion Program." *Journal of Clinical Psychology* 69, no. 1 (January 2013). https://doi.org/10.1002/jclp.21923.

Ngan Kuen Lai, Tan Fong Ang, Lip Yee Por, and Chee Sun Liew. "The Impact of Play on Child Development—A Literature Review." *European Early Childhood Education Research Journal* 26, no. 5 (September 2018): 625–43. http://dx.doi.org/10.1080/1350293X.2018.1522479.

Pascoe, Michaela C., David R. Thompson, Zoe M. Jenkins, and Chantal F. Ski. "Mindfulness Mediates the Physiological Markers of Stress: Systematic Review and Meta-Analysis." *Journal of Psychiatric Research* 95 (December 2017): 156–78. https://doi.org/10.1016/j.jpsychires.2017.08.004.

Paul, Ian M., Jennifer S. Savage, Stephanie Anzman-Frasca, Michele E. Marini, Jodi A. Mindell, and Leann L. Birch. "INSIGHT Responsive Parenting Intervention and Infant Sleep." *Pediatrics* 138, no. 1 (July 2016). https://doi.org/10.1542/peds.2016-0762.

Pearson, Jolene. "Pathways to Positive Parenting: Helping Parents Nurture Healthy Development in the Earliest Months." Washington, DC: Zero to Three, 2016.

Perry, Bruce, and Maia Szalavitz. *The Boy Who Was Raised as a Dog: And Other Stories from a Child Psychiatrist's Notebook*. New York: Basic Books, 2006.

Phillips, Lauren. "Move Over, Hygge, Gezellig Is the Trendy Danish Lifestyle Philosophy to Try." *Real Simple*, October 1, 2019. https://www.realsimple.com/work-life/life-strategies/gezellig-meaning.

Porges, Stephen. "The Infant's Sixth Sense: Awareness and Regulation of Bodily Processes." *Zero to Three* 14 (1993): 12–16. https://www.rti.org/publication/infants-sixth-sense-awareness-and-regulation-bodily-processes.

Porges, Stephen. "Neuroception: A Subconscious System for Detecting Threats and Safety." *Zero to Three* 24, no. 5 (May 2004): 19–24. https://eric.ed.gov/?id=EJ938225.

Porges, Stephen. "The Neurophysiology of Safety and How to Feel Safe." *NourishBalanceThrive* podcast, September 25, 2020. https://nourishbalancethrive.com/podcasts/nourish-balance-thrive/neurophysiology-safety-and-how-feel-safe/.

Porges, Stephen. *The Pocket Guide to the Polyvagal Theory: The Transformative Power of Feeling Safe*. New York: W. W. Norton, 2017.

Porges, Stephen. *The Polyvagal Theory: Neurophysiological Foundations of Emotions, Attachment, Communication, and Self-Regulation.* New York: W. W. Norton, 2011.

Porges, Stephen. "The Polyvagal Theory: New Insights into Adaptive Reactions of the Autonomic Nervous System." *Cleveland Clinic Journal of Medicine* 76, no. 4 suppl. 2 (February 2009): S86–S90. http://doi.org/10.3949/ccjm.76.s2.17.

Porges, Stephen. "Polyvagal Theory and Regulating Our Bodily State." Interview by D. Brown. *Affect Autism* podcast, August 24, 2020. https://affectautism.com/2020/08/24/polyvagal/.

Porges, Stephen. "Reciprocal Influences between Body and Brain in the Perception and Expression of Affect: A Polyvagal Perspective." In *The Healing Power of Emotion: Affective Neuroscience, Development, and Clinical Practice.* Edited by Diana Fosha, Daniel J. Siegel, and Marion F. Solomon, 27–54. New York: W. W. Norton, 2009.

Porges, Stephen. "Stephen Porges (Polyvagal Perspective and Sound Sensitivity Research)." The International Misophonia Research Network. Accessed May 13, 2021. https://misophonia-research.com/stephen-porges/.

Porges, Stephen, and Deb Dana, eds. *Clinical Applications of the Polyvagal Theory: The Emergence of Polyvagal-Informed Therapies.* New York: W. W. Norton, 2018.

Porges, Stephen, and Senta A. Furman. "The Early Development of the Autonomic Nervous System Provides a Neural Platform for Social Behavior: A Polyvagal Perspective." *Infant and Child Development* 20, no. 1 (February 2011): 106–18. https://doi.org/10.1002/icd.688.

Posner, Jonathan, James A. Russell, and Bradley S. Peterson. "The Circumplex Model of Affect: An Integrative Approach to Affective Neuroscience, Cognitive Development, and Psychopathology." *Development and Psychopathology* 17, no. 3 (Summer 2005): 715–34. https://doi.org/10.1017/S0954579405050340.

Price, Cynthia J., and Carole Hooven. "Interoceptive Awareness Skills for Emotion Regulation: Theory and Approach of Mindful Awareness in Body-Oriented Therapy (MABT)." *Frontiers in Psychology* 9, no. 798 (May 2018). https://doi.org/10.3389/fpsyg.2018.00798.

Raby, K. Lee, Glenn I. Roisman, R. Chris Fraley, and Jeffry A. Simpson. "The Enduring Predictive Significance of Early Maternal Sensitivity: Social and Academic Competence Through Age 32 Years." *Child Development* 86, no. 3 (May–June 2015): 695–708. https://doi.org/10.1111/cdev.12325.

Rozin, Paul, and Edward B. Royzman. "Negativity Bias, Negativity Dominance, and Contagion." *Personality and Social Psychology Review* 5, no. 4 (November 2001): 296–320. https://doi.org/10.1207/S15327957PSPR0504_2.

Russell, James A., and Lisa Feldman Barrett. "Core Affect, Prototypical Emotional Episodes, and Other Things Called Emotion: Dissecting the Elephant." *Journal of Personality and Social Psychology* 76, no. 5 (May 1999): 805–19. https://doi.org/10.1037//0022-3514.76.5.805.

Salovey, Peter, and John D. Mayer. "Emotional Intelligence." *Imagination, Cognition and Personality* 9, no. 3 (March 1990): 185–211. https://doi.org/10.2190%2FDUGG-P24E-52WK-6CDG.

Savage, Jennifer S., Emily E. Hohman, Michele E. Marini, Amy Shelly, Ian M. Paul, and Leann L. Birch. "INSIGHT Responsive Parenting Intervention and Infant Feeding Practices: Randomized Clinical Trial." *International Journal of Behavioral Nutrition and Physical Activity* 15, no. 64 (July 2018). https://doi.org/10.1186/s12966-018-0700-6.

Schaaf, Roseann C., and Lucy Jane Miller. "Occupational Therapy Using a Sensory Integrative Approach for Children with Developmental Disabilities." *Mental Retardation and Developmental Disabilities Research Reviews* 11, no. 2 (April 2005): 143–48. https://doi.org/10.1002/mrdd.20067.

Schnabel, Alexandra, David J. Hallford, Michelle Stewart, Jane A. McGillivray, David Forbes, and David W. Austin. "An Initial Examination of Post-Traumatic Stress Disorder in Mothers of Children with Autism Spectrum Disorder: Challenging Child Behaviors as Criterion A Traumatic Stressors." *Autism Research* 13, no. 9 (September 2020): 1527–36. https://doi.org/10.1002/aur.2301.

Seligman, Martin. *Flourish: A Visionary New Understanding of Happiness and Well-Being*. New York: Simon & Schuster, 2011.

Seligman, Martin, Tracy A. Steen, Nansook Park, and Christopher Peterson. "Positive Psychology Progress: Empirical Validation of Interventions." *American Psychologist* 60, no. 5 (July–August 2005): 410–21. https://doi.org/10.1037/0003-066X.60.5.410.

Shanker, Stuart. *Reframed: Self-Reg for a Just Society*. Toronto: University of Toronto Press, 2020.

Shonkoff, Jack P., and Deborah A. Phillips, eds. *From Neurons to Neighborhoods: The Science of Early Childhood Development*. Washington, DC: The National Academies Press, 2000.

Siegel, Daniel. *The Mindful Brain: Reflection and Attunement in the Cultivation of Well-Being*. New York: W. W. Norton, 2007.

Siegel, Daniel, and Tina Payne Bryson. *The Whole-Brain Child: 12 Revolutionary Strategies to Nurture Your Child's Developing Mind*. New York: Random House, 2011.

Siegel, Daniel, and Mary Hartzell. *Parenting from the Inside Out: How a Deeper Self-Understanding Can Help You Raise Children Who Thrive*. New York: Penguin, 2004.

Southwick, Steven, and Dennis Charney. *Resilience: The Science of Mastering Life's Greatest Challenges*. Cambridge: Cambridge University Press, 2018.

Stanfield, Cindy. *Principles of Human Physiology*. 4th ed. San Francisco: Pearson Education, 2011.

Stanford Medicine, The Center for Compassion and Altruism Research and Education. "Emotion Resonance." Accessed May 18, 2021. http://ccare.stanford.edu/research/wiki/compassion-definitions/emotion-resonance/.

Steiner, Claude. In "What Is Emotional Literacy?" Habits for Wellbeing. Accessed May 18, 2021. https://www.habitsforwellbeing.com/what-is-emotional-literacy/.

Streeter, Chris C., Patricia L. Gerbarg, Theodore H. Whitfield, Liz Owen, Jennifer Johnston, Marisa M. Silveri, Marysia Gensler, et al. "Treatment of Major Depressive Disorder with Iyengar Yoga and Coherent Breathing: A Randomized Controlled Dosing Study." *Journal of Alternative and Complementary Medicine* 23, no. 3 (March 2017): 201–7. https://doi.org/10.1089/acm.2016.0140.

Taj-Eldin, Mohammed, Christian Ryan, Brendan O'Flynn, and Paul Galvin. "A Review of Wearable Solutions for Physiological and Emotional Monitoring for Use by People with Autism Spectrum Disorder and Their Caregivers." *Sensors* 18, no. 12 (December 2018): 4271. https://doi.org/10.3390/s18124271.

Task Force on Sudden Infant Death Syndrome. "SIDS and Other Sleep-Related Infant Deaths: Updated 2016 Recommendations for a Safe Infant Sleeping Environment." *Pediatrics* 138, no. 5 (November 2016). https://doi.org/10.1542/peds.2016-2938.

Tronick, Edward. "Emotions and Emotional Communication in Infants." *American Psychologist* 44, no. 2 (February 1989): 112–19. https://doi.org/10.1037//0003-066x.44.2.112.

Tronick, Edward. *The Neurobehavioral and Social-Emotional Development of Infants and Children*. New York: W. W. Norton, 2007.

Tronick, Edward, and Marjorie Beeghly. "Infants' Meaning-Making and the Development of Mental Health Problems." *American Psychologist* 66, no. 2 (February–March 2011): 107–19. https://doi.org/10.1037/a0021631.

Twal, Waleed O., Amy E. Wahlquist, and Sundaravadivel Balasubramanian. "Yogic Breathing When Compared to Attention Control Reduces the Levels of Pro-Inflammatory Biomarkers in Saliva: A Pilot Randomized Controlled Trial." *BMC Complementary and Alternative Medicine* 16, no. 294 (August 2016). https://doi.org/10.1186/s12906-016-1286-7.

United States Department of Education Office for Civil Rights. "Data Snapshot: Early Childhood Education." Issue Brief No. 2, March 2014. https://www2.ed.gov/about/offices/list/ocr/docs/crdc-early-learning-snapshot.pdf.

Van der Kolk, Bessel. *The Body Keeps the Score: Brain, Mind, and Body in the Healing of Trauma.* New York: Penguin Books, 2014.

Vygotsky, Lev S. *Mind in Society: The Development of Higher Psychological Processes.* Edited by Michael Cole, Vera John-Steiner, Sylvia Scribner, and Ellen Souberman. Cambridge, MA: Harvard University Press, 1978.

Walker, Matthew. "Sleep Is Your Superpower." Filmed April 2019 at TED conference. https://www.ted.com/talks/matt_walker_sleep_is_your_superpower.

Walker, Matthew. *Why We Sleep: Unlocking the Power of Sleep and Dreams.* New York: Scribner, 2017.

Waters, Sara F., Tessa V. West, and Wendy Berry Mendes. "Stress Contagion: Physiological Covariation between Mothers and Infants." *Psychological Science* 25, no. 4 (April 2014): 934–42. https://doi.org/10.1177/0956797613518352.

Wiking, Meik. *The Little Book of Hygge: Danish Secrets to Happy Living.* New York: HarperCollins, 2017.

Winnicott, Donald W. "Mirror-Role of Mother and Family in Child Development." In *Playing and Reality.* 2nd ed., 149–59. London: Routledge Classics, 2005.

Wolpe, Joseph. *The Practice of Behavior Therapy.* London: Pergamon Press, 1973.

Yogman, Michael, Andrew Garner, Jeffrey Hutchinson, Kathy Hirsh-Pasek, Roberta Michnick Golinkoff, Committee on Psychosocial Aspects of Child and Family Health, and Council on Communications and Media. "The Power of Play: A Pediatric Role in Enhancing Development in Young Children." *Pediatrics* 142, no. 3 (September 2018): 1–17. https://psycnet.apa.org/record/2018-54541-014.

Zero to Three. "Judgment." https://www.zerotothree.org/resources/series/judgment.

Zero to Three. "National Parent Survey Overview and Key Insights." June 6, 2016. https://www.zerotothree.org/resources/1424-national-parent-survey-overview-and-key-insights.

Zero to Three. "Parent Survey Reveals Expectation Gap for Parents of Young Children." Updated October 13, 2016. https://www.zerotothree.org/resources/1612-parent-survey-reveals-expectation-gap-for-parents-of-young-children.

Zero to Three. "Responsive Care: Nurturing a Strong Attachment through Everyday Moments." February 22, 2016. https://www.zerotothree.org/resources/230-responsive-care-nurturing-a-strong-attachment-through-everyday-moments.

Zessin, Ulli, Oliver Dickhäuser, and Sven Garbade. "The Relationship between Self-Compassion and Well-Being: A Meta-Analysis." *Applied Psychology: Health and Well-Being* 7, no. 3 (November 2015): 340–64. https://doi.org/10.1111/aphw.12051.

INDEX

ABOUT THE AUTHOR

MONA DELAHOOKE, PH.D., is a licensed clinical psychologist and mother who has cared for children and their families for more than three decades. She is a frequent speaker, trainer, and consultant to parents, organizations, schools, and public agencies. Dr. Delahooke has dedicated her career to promoting compassionate, relationship-affirming practices in parenting, education, and psychology by translating current neuroscience into practical tools and strategies.

Her award-winning book *Beyond Behaviors: Using Brain Science and Compassion to Understand and Solve Children's Behavioral Challenges* called for a paradigm shift in education and psychology, from focusing on children's behaviors to the reasons underlying behaviors. It provided a new framework for supporting children with behavioral challenges. Building on those lessons, *Brain-Body Parenting* adds the important insight that the brain-body connection provides a new foundation for helping parents raise joyful and resilient kids.

Dr. Delahooke's popular blog, www.monadelahooke.com, covers a range of topics useful for parents and all childhood providers. You can find Mona on Facebook, Instagram, and Twitter @monadelahooke.